D1806088

M & E HANDBOOKS

M & E Handbooks are recommended reading for examination syllabuses all over the world. Because each Handbook covers its subject clearly and concisely books in the series form a vital part of many college, university, school and home study courses.

Handbooks contain detailed information stripped of unnecessary padding, making each title a comprehensive self-tuition course. They are amplified with numerous self-testing questions in the form of Progress Tests at the end of each chapter, each text-referenced for easy checking. Every Handbook closes with an appendix which advises on examination technique. For all these reasons, Handbooks are ideal for pre-examination revision.

The handy pocket-book size and competitive price make Handbooks the perfect choice for anyone who wants to grasp the essentials of a subject quickly and easily.

THE M. & E. HANDBOOK SERIES

HUMAN AND SOCIAL BIOLOGY

GEORGE USHER, B.Sc., Dip.Agric.Sci., D.T.A., M.I.Biol., F.L.S.

Senior Biology Master,
Bedstone School, Bucknell, Salop

MACDONALD AND EVANS

MACDONALD AND EVANS LTD.
Estover, Plymouth PL6 7PZ

First published 1977
Reprinted in this format 1978

©

MACDONALD AND EVANS LIMITED
1977

ISBN: 0 7121 0808 4

*Printed in Great Britain by Richard Clay (The Chaucer Press), Ltd.,
Bungay, Suffolk*

PREFACE

THE emphases in the study of human and social biology have
radically altered during the last few years. This HANDBOOK
has been written to meet the new requirements, with extensions
where necessary to make it suitable for overseas students as well
as those studying in the United Kingdom. The syllabuses are
divided into three related units, namely: man—the individual,
man in relation to other organisms, and man in the community.
For ease of studying I have divided this book into much shorter
chapters, but this overall sequence is followed.

Examination questions are of three types—the normal essay,
structured answers, and multiple-choice—and the essay questions
are not necessarily confined to the "one topic—one question" type
as they used to be. In spite of this the questions at the end of the
chapters are limited to materials found in the chapter. This has
been done so that answering the questions can be a reinforcement
to learning the material rather than a test in examination technique. If facts are thoroughly learnt, their application becomes
easy.

Only a few experiments are described, as they would be out of
place in a concise HANDBOOK of this type, but it should be remembered that experiment is a part of any science, and they
should be performed whenever possible.

October, 1976   G. U.

CONTENTS

CHAPTER I

CELLS AND TISSUES

1. Introduction. All living things are made up of one or more cells. Some plants and animals are single-celled, but the human body contains thousands of millions of cells. These cells are differentiated into *tissues*, which themselves are made up into *organs*. Each tissue contains cells of one type which perform a particular function, *e.g.* nerve tissue is made up of cells which are specialised to transmit impulses. Organs contain different tissues, *e.g.* the heart contains muscle tissue, nerve tissue, connective tissue, etc.

CELLS

2. The typical cell. Each cell is a mass of living substance called *protoplasm*. It is surrounded by a fine membrane called the *plasma membrane* and within it is the *nucleus* which controls

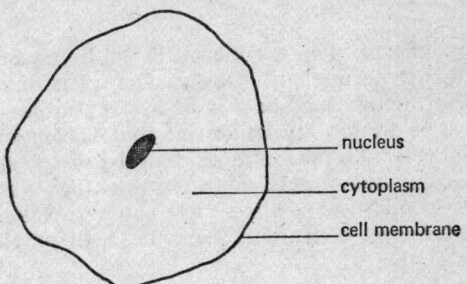

Fig. 1. A typical cell.

the activities of the cell. The epithelial cells from the inside of the cheek can be used to illustrate these points. Firmly scrape the inside of the cheek with a sterilised instrument and mount the scrapings on a microscope slide. Look at the slide under the microscope. Prepared slides may be available which will be stained to show the nucleus more clearly (*see* Fig. 1).

3. The plasma membrane. Any materials, *e.g.* food and oxygen, entering the cell have to pass through the plasma membrane; similarly substances leaving the cell, *e.g.* carbon dioxide and excretory products, also have to pass through this membrane. Some of these substances diffuse through the

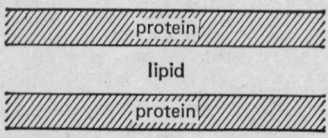

FIG. 2. A plasma membrane.

membrane, while others are "pumped" through it by the process of active transport. This process involves the use of energy, and stops if the cell stops respiring. A similar membrane surrounds the nucleus, and regulates the flow of materials between the nucleus and the cytoplasm. These membranes are 70–100 Å wide (*see* Fig. 2).

4. The cytoplasm. The cytoplasm is the living substance of the cell, except the nucleus. Observation of it under the low power of the microscope show it as a structureless, jelly-like mass; but it is within this material that the metabolic processes of the cell take place. More detailed observations using special staining techniques, high-powered light microscopes or an electron microscope show that the cytoplasm is a very complex substance, containing various small particles called "organelles".

(*a*) *Endoplasmic reticulum* (*Cytoplasmic reticulum*). This is a complex network which occupies the greater part of the cytoplasm. Structurally it is similar to the plasma membrane and has the effect of increasing the surface within the cell. The large surface area is necessary for enzyme actions. The network may also divide the cell into separate zones in which different physiological reactions can take place.

Minute particles (*ribosomes*) concerned with the synthesis of proteins are attached to the membrane.

(*b*) *Mitochondria*. These are approximately 0.5μ in width and vary in length, but are usually $1-3\mu$ long. Electron microscope studies reveal that they are hollow, the wall

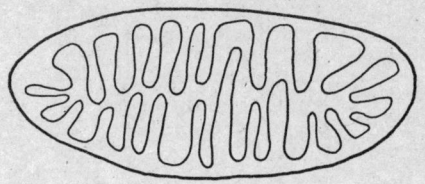

Fig. 3. Section through a mitochondrion.

being structurally similar to the plasma membrane. The inner layer of the wall is extended into ridges into the vacuole. This has the effect of increasing the surface area. The mitochondria are sometimes called the "power houses" of the cell because they are concerned with aerobic respiration (*see* II). Each cell contains many thousand mitochondria (*see* Fig. 3).

5. The nucleus.

(*a*) *Structure*. The nucleus is usually spherical, its size depending on the overall volume of the cell, and the amount of physiological activity which takes place in the cell. It is surrounded by a membrane (*nuclear membrane*) which is the extension of the endoplasmic reticulum, and in which there are pores, so that there is a free interchange of materials between the nucleus and the surrounding cytoplasm. In prepared slides, the nucleus stains more darkly than the surrounding cytoplasm due to the presence of a substance called "chromatin".

(*b*) *Function*. The nucleus is the control centre of the cell. To bring about this control it contains *genes*, which are located on variously-shaped *chromosomes*. Chromosomes are not visible in the normal cell, but become evident if the cells are stained when they are dividing. The presence of genes is inferred from their activity, and they are assumed to be like a string of beads arranged on the chromosome. (We will see the implications of this when we discuss "Heredity" in VIII.)

(c) *Chromosomes*. Each chromosome contains a twisted double strand of deoxyribosenucleic acid (DNA), so that the DNA is rather like a twisted ladder. Along the uprights of

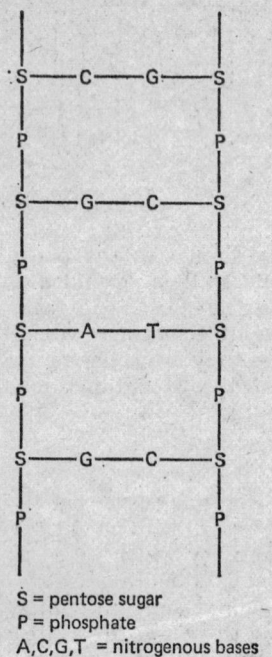

S = pentose sugar
P = phosphate
A,C,G,T = nitrogenous bases

FIG. 4. A small piece of DNA.

the ladder are arranged four nitrogenous bases (adenine, cytosine, guanine, and thymine) in different numbers and orders. It is these small groups of nitrogenous bases which constitute the genes. When the cell divides the DNA strands split longitudinally, one half passing into each new cell where it replicates the missing half. This means that each cell has the same DNA complement (*see* Fig. 4).

(d) *Functions of the genes.* Each gene is made up of at least three pairs of the nitrogenous bases mentioned above. The gene acts as a template for the production of ribose nucleic acid (RNA) which contains similar constituents to DNA, except that the thymine is replaced by uracil. When a particular RNA grouping is formed, it passes from the nucleus to the ribosomes where proteins are synthesised. Here a particular enzyme is synthesised, all enzymes being proteins. The enzyme will then take part in a particular chemical reaction. Thus the genes in the nucleus govern the reactions of the cell. For reasons not fully understood, different genes function in different cells, for, as we shall see, every cell in the body (except sex cells) contains the same gene compliment.

6. Cell division. A human being begins life as a single cell (*zygote*) which results from the fusion of a sperm cell (*spermatozoon*) with an egg cell (*ovum*). This cell then divides to form an embryo. Further divisions take place as the child grows until it is fully grown. During the growth of an individual, the cells *differentiate* into different tissues, each performing their different functions in the fully-developed individual.

Further, the fully-developed tissues are constantly "wearing out", and their cells have to be replaced, so that throughout the life of an individual new cells are being constantly formed by the division of the old ones. The process of cell division is called *mitosis*.

7. Mitosis. During this process, the cell divides to produce two identical daughter cells. The DNA splits longitudinally as described, so that each daughter cell has the same DNA compliment in its nucleus.

Although mitosis is a continuous process, it is divided for convenience, into four stages. They are called *prophase*, *metaphase*, *anaphase*, and *telophase*. These stages can be seen in prepared slides of Broad Bean root tips, or the later stages of the development of Roundworm eggs. Human cells are usually too small, or the nuclei are too obscured to see mitosis clearly.

(a) *Prophase.* The chromosomes become readily stainable, and are seen as paired, coiled threads (*chromatids*) joined

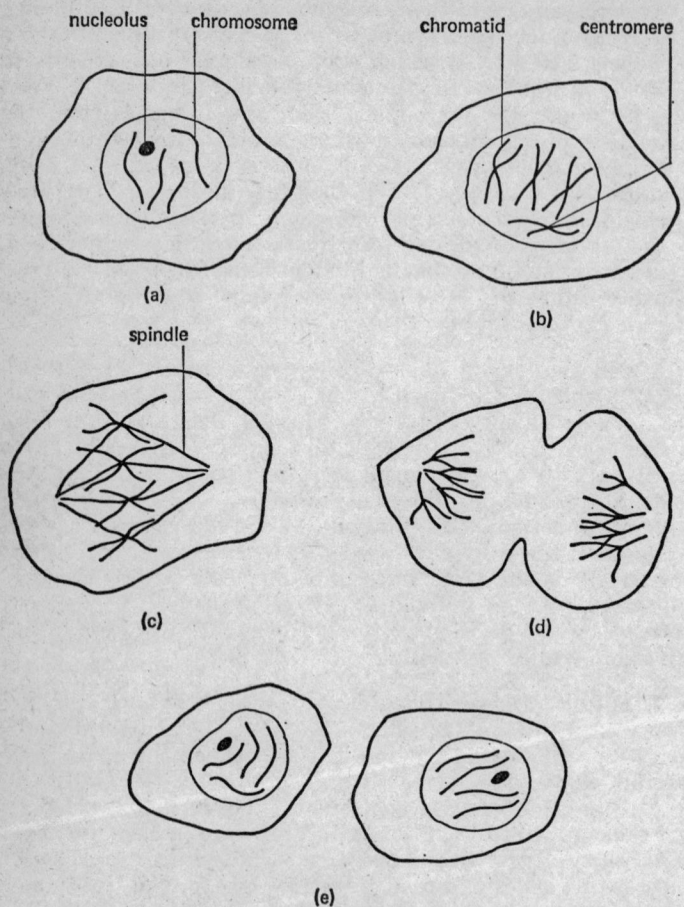

Fig. 5. Mitosis: (a) interphase; (b) early prophase; (c) metaphase; (d) anaphase; (e) end of telophase.

together near the centre by a constriction (the *centromere*). As prophase proceeds the chromatids become shorter and thicker. A small body outside the nucleus (the *centriole*) divides and the halves migrate to opposite sides (poles) of the nucleus. The nuclear membrane disappears.

(b) *Metaphase.* The chromosomes become arranged centrally as a plate across the cell, and attached to the *spindle* by their centromeres. The spindle is formed between the two halves of the centriole, which are at the opposite poles of the cell by this time. The centromeres split longitudinally.

(c) *Anaphase.* The spindle fibres contract and the separate chromatids move to opposite ends of the cell.

(d) *Telophase.* The chromosomes become less readily stained. The spindle fibres disappear and nuclear membranes are formed around the two daughter nuclei.

At the beginning of prophase a relatively large body within the nucleus (*nucleolus*) disappears, and is reformed in each new nucleus during telophase (*see* Fig. 5).

TISSUES

8. Introduction. The single-celled zygote divides into 2, 4, 8, then 16, etc. cells, and soon forms a small ball of cells from which the embryo develops. The cells grow and divide at different rates, and soon various groups of them become differentiated into tissues. The tissues mentioned below should be examined under the microscope.

9. Striated muscle. This tissue makes up the bulk of the flesh of a human being. It is capable of contracting, so that the whole muscle can move the bones (*see* IV). Each muscle fibre is made up of alternate light and dark bands, and the

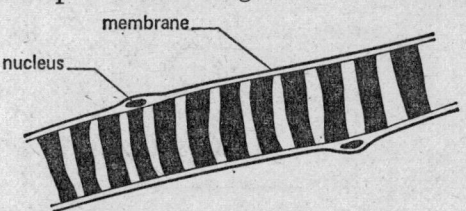

Fig. 6. Fibre of striated muscle.

whole is surrounded by a fine membrane. Cell nuclei are seen underneath the membrane. When the muscle contracts, the bands are pulled closer together, making the fibre shorter and thicker. This process requires energy, which is derived from respiration (*see* Fig. 6).

Haversian canal　　　　　　osteocyte

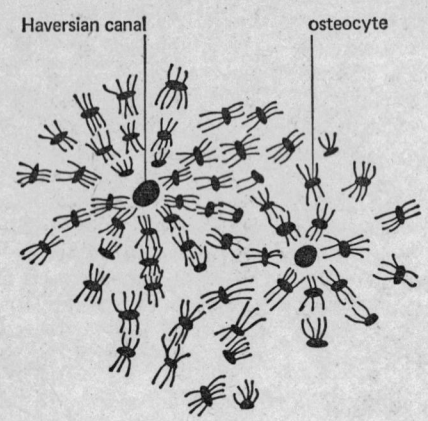

FIG. 7. Bone.

perichondrium

matrix

cells

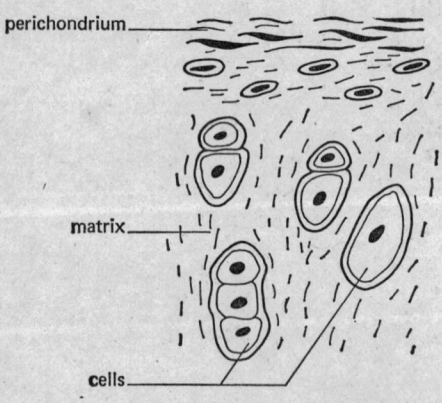

FIG. 8. Cartilage.

10. Bone. The bulk of bone is calcium phosphate which is secreted by cells called *osteocytes* (*see* Fig. 7). Connective tissue is also present. Within this matrix are fine canals through which blood vessels and nerves run (*see* IV).

11. Cartilage. There are various types of cartilage, some are elastic, and others are not. It is found between bones as a "shock-absorbing" and lubricating tissue, and as a supporting tissue in such places as the nose and ear. The greater part of cartilage is a tough, elastic substance which is secreted by cells embedded in it (*see* Fig. 8).

12. Nerve cells (Neurones). The nerve cell consists of a *cell body* which contains the nucleus, and various extensions from it. The longest extension is called the *axon*, which may be over a metre long in some of the spinal nerves, while the shorter ones are *dendrons*. There are finer branches at the ends of these primary extensions called "dendrites". The

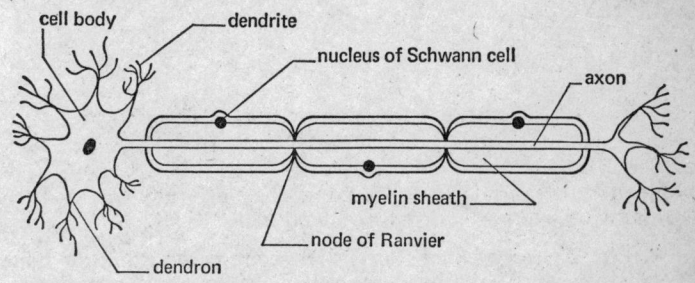

FIG. 9. Motor neurone.

dendrites from adjoining nerve cells intermingle to form a *synapse*. By this arrangement nerve impulses can pass from one part of the body to another very quickly (*see* V). The axon is surrounded by a sheath of fatty material (*myelin sheath*) which is interrupted at intervals by the *nodes of Ranvier*. Nerve impulses jump from node to node, and so their passage along the axon is accelerated (*see* Fig. 9).

13. Blood. (*See* III and VII.) Blood is a fluid tissue made up of the blood *plasma* in which are suspended the red and white cells and small non-nucleate platelets.

(*a*) *Red cells.* These are produced in the marrow of the long bones and ribs. They are biconcave discs about 7μ in diameter. No nucleus is present. The red colour is due to the pigment *haemoglobin* which is responsible for the transport of oxygen. There are about five million red blood cells (*erythrocytes*) in 1 (cubic millimetre) (mm^3) of blood.

(*b*) *White blood cells.* The white blood cells are slightly larger than erythrocytes. There are fewer of them than red

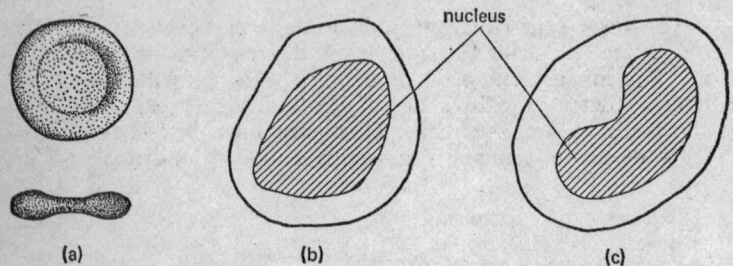

Fig. 10. Blood cells: (*a*) erythrocytes; (*b*) lymphocyte; (*c*) leucocyte.

blood cells (about 1 to 500), and they are nucleate. There are two basic types of white blood cells.

(*i*) *Leucocytes* (*monocytes*). These are produced in the bone marrow and have bean-shaped nuclei. They ingest foreign proteins and bacteria which have been inactivated by antibodies.

(*ii*) *Lymphocytes.* These are produced in the lymphatic tissue and secrete antibodies. They are capable of migrating through the walls of the blood vessels, and so can come into intimate contact with cells anywhere in the body.

(*c*) *Platelets.* Platelets are less than half the size of the red blood cells from which they may be derived. No nucleus is present. They are concerned with the clotting of the blood.

(d) *Plasma*. This pale yellow liquid makes up about 55 per cent of the volume of the blood. It contains the dissolved substances which are transported by the blood, including the protein *fibrin* which assists in clotting. When the fibrin is removed artificially by allowing the blood to clot, the resulting liquid is *serum*. Do not confuse the terms plasma and serum (*see* Fig. 10).

14. Surfaces. It is well to remember at this point that materials have to enter tissues through their external surfaces. In all organs there are various structures which increase the surface without increasing the volume of the cells, so that materials can enter the organs more quickly, *e.g.* the walls of the intestine are thrown up into *villi*, the lungs are divided into *alveoli*. Similarly the capillaries of the blood system are very thin-walled and ramify through the tissues, so that no cell is any great distance from the blood supply which is bringing it essential food supplies and oxygen.

PROGRESS TEST 1

1. Distinguish between cells, tissues and organs.   **(1)**
2. What are the three main parts of a cell?   **(2)**
3. What is the main function of the plasma membrane?   **(3)**
4. What are organelles?   **(4)**
5. State the function of the endoplasmic reticulum.   **(4)**
6. What is the function of the ribosomes?   **(4)**
7. Draw a mitochondrion.   **(4)**
8. What is the function of the mitochondria?   **(4)**
9. Where is DNA found, and what do the letters stand for? **(5)**
10. What are the functions of genes?   **(5)**
11. State the stages in mitosis.   **(7)**
12. Draw and label a piece of striated muscle.   **(9)**
13. What is the function of cartilage in a joint?   **(11)**
14. Draw a neurone. **(12)**
15. Name the constituents of the blood.   **(13)**

EXAMINATION QUESTIONS

1. Draw a diagram of an animal cell, *e.g.* from inside the cheek. Label the nucleus, plasma membrane, a mitochondrion, a ribosome and the endoplasmic reticulum. Describe the functions of the parts you have labelled.

2. Write about ten lines on the following: (a) DNA; (b) genes; (c) chromosomes.

3. Describe the process of mitosis. Why is this method of cell division important?

4. Describe the following tissues: (a) striated muscle; (b) cartilage; (c) a neurone. What parts do these tissues play in the movement of a joint?

5. Describe the structures found in the blood. What are their functions?

CHAPTER II

FOOD, NUTRITION AND ENERGY

INTRODUCTION

1. Types of nutrition. All animals, including man, are dependent on plants for their food. Green plants produce their food by *photosynthesis*, whereby carbon dioxide and water are combined together by light energy to form glucose. This process needs the green pigment chlorophyll.

$$6CO_2 + 6H_2O \rightarrow C_6H_{12}O_6 + 6O_2$$

Carbon Dioxide   Water   Glucose   Oxygen

In the green plant, some of the intermediates of the photosynthetic process are combined with nitrogen-containing radicals from the soil to produce proteins.

Thus green plants utilise simple compounds to make more complex ones, and in doing so they convert light energy from the sun into chemical energy, and incidentally replace the carbon dioxide produced by plants and animals by oxygen.

Animals feed on plants or other animals. In doing so, they *digest* the complex substances in their food. The smaller units produced by digestion are used by the animals to build up their own tissues, *i.e.* they are used during the production of new tissues during growth, or to replace old cells during repair. Some of the small molecules produced during digestion are broken down further in the cells to release energy. This process is called *respiration*.

All the materials that are used during the nutrition of plants and animals are returned ultimately into the environment so that the elements that they contain can be recirculated. If this did not happen, all the raw materials would have been used up a long time ago. The recirculation of elements in nature can be represented by cycles, illustrated by the carbon cycle.

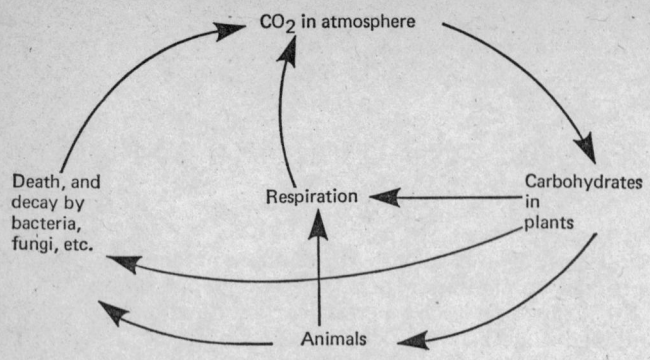

The carbon cycle

Similar cycles can be constructed for the other elements. Notice the important part played by bacteria (and other organisms) in the release of simple compounds for recirculation.

2. Food chains. The dependence of man on plants for his food can be shown by constructing food chains. Below are two examples, but you can construct your own.

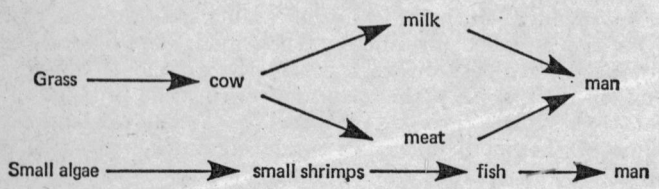

Food chains

At each stage in the food chain, some of the sun's energy fixed by the green plant, is used up by the animal eating the plant, so that less is left to pass on to the next animal in the chain. Therefore the food reserves of the world would be better conserved if man lived directly on plants, *i.e.* was entirely vegetarian. As the food situation in the world becomes more critical, more attention is being paid to the direct conversion of plants into forms acceptable to human beings. For example, synthetic meat is being made entirely from plant protein.

3. Calorific value. The energy obtained from food is expressed in Calories (spelt with a capital "C"). This is a unit of heat energy which, although no longer used in other branches of science, is still used when referring to foods. The calorie (spelt with a small "c") is the amount of heat required to raise the temperature of 1 gram (g) of water from 17°C to 18°C. The Calorie is 1,000 calories.

The Calorific value of a food is the amount of heat which can be obtained from 1g of the food by complete combustion. The Calorific values for carbohydrates and proteins are approximately 4·1 Cal/g and for fats 9·3 Cal/g. To put this into everyday terms, it would take the heat from about 7 g of carbohydrate to boil half-a-pint of freezing water.

TYPES AND SOURCES OF FOODS

4. Introduction. There are six types of foods, all of which are needed in the diet of a human being. They are carbohydrates, fats, proteins, water, minerals, and vitamins.

5. Carbohydrates. These are chemical compounds containing only carbon, hydrogen, and oxygen, the last two elements being in the same proportions as they are in water. Carbohydrates have the general formula $C_xH_{2y}O_y$. Some of the simplest carbohydrates occurring in foods are the *hexose* or *monosaccharide* sugars which have the general formula $C_6H_{12}O_6$. Glucose is the most important of these, but fructose is also important as it occurs in fruits. The *disaccharides* are carbohydrates with the general formula $C_{12}H_{22}O_{11}$. Sucrose (cane, or beet sugar) is the most important example of these, but maltose (malt sugar) and lactose (milk sugar) also play a prominent part in the nutrition of a human being. Carbohydrates with larger molecules than these are called *polysaccharides* which include starch, glycogen, and cellulose.

Starch is the chief source of polysaccharides used by man. Glycogen is a starch-like compound forming a carbohydrate store in the liver and muscles. Cellulose forms the cell walls of green plants, and so large quantities are taken in with the food by human beings. Man is incapable of digesting cellulose, so its only importance in the diet is to add bulk to the food, and in doing so it aids bowel movement and prevents constipation.

Carbohydrates are the chief energy-producing foods, yielding 4·1 Cal/g. As will be seen later, they are broken down completely during respiration into carbon dioxide and water.

Carbohydrates that are not required immediately as a source of energy are stored in the liver and muscles as glycogen, or are stored for longer periods as fat laid down under the skin and around organs, especially the kidneys.

The monosaccharide sugars are particularly important, as it is in this form only that carbohydrates can be absorbed directly through the intestine. This is why glucose is taken by athletes and invalids to give a quick and easily assimilated source of energy. During digestion other forms of carbohydrate are *hydrolysed* to monosaccharides by the addition of molecules of water. This process is brought about by enzymes, *e.g.*:

$$\underset{\text{sucrose}}{C_{12}H_{22}O_{11}} + \underset{\text{water}}{H_2O} \rightarrow \underset{\text{glucose}}{C_6H_{12}O_6} + \underset{\text{fructose}}{C_6H_{12}O_6}$$

The presence of glucose in a food can be shown by warming it with Fehling's solution, or Benedict's solution. An orange-to-red precipitate confirms the presence of glucose.

Hydrolysis of sucrose can be demonstrated by the following experiment:

Dissolve a little sucrose in water, and divide the solution into two. Test one portion with Fehling's or Benedict's solution. This should give a negative result. Boil the other portion with a little dilute hydrochloric acid which hydrolyses the sucrose. Make the solution alkaline by adding sodium hydroxide solution, and test with Fehling's or Benedict's solution. A positive result should now be obtained.

The presence of starch in a food is shown by adding a few drops of iodine solution (in potassium iodide solution or ethanol). A blue-black colouration indicates the presence of starch.

Sources of carbohydrates:

(a) *Starch*—wheat, rice, corn (maize), millet, sorghum, potatoes, yams, taro, sweet potato, plantains.

(b) *Sucrose*—sugar cane, sugar beet.

(c) *Glucose*—dates, grapes, carrots, bananas.

(d) *Fructose*—fruits, honey.

(e) *Glycogen*—liver, lean meat.

(f) *Lactose*—milk and whole milk products, *e.g.* cheese and yogurt.

6. Fats. These compounds contain carbon, hydrogen and oxygen, but there is a smaller proportion of oxygen than in carbohydrates. Fats that are liquid at about 15°C are called *oils*. They are insoluble in water, so that during digestion they have to be hydrolysed into acids (so-called fatty acids) and glycerol, both of which are water-soluble. Fats produce over twice the amount of energy produced by carbohydrates (9·3 Cal/g) but their main function is as an "energy store" being converted to carbohydrates when needed. Deposits of fat laid down under the skin act as an insulator retaining the body heat. They also are solvent for some vitamins.

When fats are oxidized they produce carbon dioxide and water.

Sources of fats:

(a) *Animal fats*—butter, lard, fat meat, whale oil, fish liver.

(b) *Plant fats* (*oils*)—ground nuts (peanuts), coconut oil, palm fruit, sunflower seed, cotton seed. Many seeds and fruits contain appreciable amounts of oils.

Whale oil and the plant oil mentioned are used in the manufacture of margarine.

(c) Milk contains 3–4 per cent fat.

The presence of fat in a food can be demonstrated by grinding a little of the food in water and adding the stain Sudan III. This stains the fat pink, and the pink droplets float to the surface.

7. Proteins. The proteins contain the elements carbon, hydrogen, oxygen and *nitrogen*; sulphur and other elements are often present as well. Protein molecules are very large, containing tens of thousands of atoms. In spite of their large size they are made up of twenty amino acids. These amino acids are linked together in different numbers and combinations to give a very large number of different proteins. There are many different amino acids, but there are only these twenty *essential amino acids* needed to synthesise all the proteins in the human body. It is important that foods contain the proteins which are made up of these twenty amino acids—any others are useless. First-class proteins contain all the essential amino acids, while second-class proteins contain only some of them.

Protein molecules are too large to be absorbed directly and have to be hydrolysed during digestion into their constituent amino acids which can be absorbed.

The protoplasm of cells contains a large amount of protein, so proteins are particularly important in the diet of growing children who are producing a lot of new cells. Even in the adult, the cells are continually dying and have to be replaced (red blood cells are a particularly good example of this), so protein is needed in their diet also.

Proteins can be respired to produce a similar amount of energy as carbohydrates. The products of protein respiration are carbon dioxide, water and urea. The human body cannot store protein, so that a continuous supply is needed. Any excess over immediate requirements is removed by the liver as urea.

Sources of proteins:

(a) *First-class protein*—meat (any kind) (about 15 per cent protein), fish, eggs, cheese, milk (contains about 3·5 per cent protein).

(b) *Second-class protein*—peas, beans, lentils, soy bean, ground nuts, wheat.

The presence of protein can be demonstrated by boiling a portion of the food with Millon's reagent which colours the protein pink.

8. Water. Although water has no food value, it is an essential part of the diet. Cells contain large amounts of water, and the human body as a whole is about 70 per cent water. As well as being a constituent of protoplasm, water is the chief solvent of foods and other substances entering and leaving the cells. Cells can only take in or release substances which are dissolved in water. The evaporation of water from the skin helps to keep the body cool. Water is also lost from the lungs during breathing, as urine during the removal of excretory substances, and in the faeces. It is replaced by drinking, in foods with a high water content, e.g. vegetables, and by the respiration of carbohydrates, fats and protein. This latter is called metabolic water.

9. Mineral salts. A large number of mineral salts are needed by the body. Some, e.g. calcium, are needed in relatively large amounts, while others, e.g. copper, are needed in very small amounts. These latter are called *trace elements*.

(a) *Calcium*—needed to produce bones and teeth. Bone is made largely of calcium phosphate. Calcium is also needed to clot the blood. It is particularly important in the diets of growing children and expectant mothers. If an expectant mother has a calcium deficient diet, the calcium in her bones is mobilised to provide the needs of her developing child.

Deficiency in young children causes a malformation of the bones called "rickets" (*see below*—vitamin D).

Milk, cheese, eggs, and green vegetables are good sources of calcium.

(b) *Phosphorus*—also needed for bone formation. It is also an important constituent of the protein in cell nuclei, and is therefore necessary for cell division.

Phosphorus is found in most protein-containing foods, especially milk and cheese. Vegetables are also a fairly good source.

(c) *Potassium*—required for the normal production of protoplasm by cells, and to keep the constitution of the blood plasma constant.

It is found in green vegetables and meats.

(d) *Sulphur*—an essential constituent of protoplasm.

Any protein-containing food will provide sufficient of this element.

(e) *Chlorine*—an important element in maintaining the constitution of the blood plasma. It is also needed to produce hydrochloric acid in the stomach.

Common salt (sodium chloride) is an excellent source.

(f) *Sodium*—also needed to keep the constitution of the blood plasma constant. It is also necessary for the transport of carbon dioxide in the blood.

Common salt is an excellent source.

(g) *Magnesium*—needed for the formation of teeth and bones.

It is found in sufficient amounts in most foods.

(h) *Fluorine*—necessary to harden the enamel of the teeth, and so reduce the incidence of decay.

Fluorine occurs naturally in drinking water, but in some areas the addition of fluorides to the drinking water has reduced the incidence of dental caries (tooth decay), particularly among young children.

(i) *Iodine*—needed for the production of the hormone thyroglobulin (thyroxine) which controls growth. Deficiency of the hormone leads to mental and physical retardation in children, called "cretinism", and "myxoedema" in adults. The symptoms of myxoedema are a drying of the skin and hair, swelling of the feet and hands, a slowing of the mental processes and irritability. Thyroglobulin is produced in the thyroid glands. A deficiency of iodine in the diet causes these glands (in the neck) to become overactive and enlarge to form a goitre.

Sufficient iodine is usually found in the drinking water, but fish and seafoods (particularly seaweeds) are good sources.

(j) *Iron*—a constituent of haemoglobin. It is particularly important in the diet of expectant mothers and adolescents. Deficiency of iron leads to anaemia, which is characterised by a general tiredness and paleness.

Meat (especially liver) and green vegetables (especially spinach and water cress) are good sources of iron.

(k) *Copper and cobalt*—both are needed for the formation of haemoglobin and are found as trace elements in vegetables.

(l) *Zinc*—necessary for the production of the hormone insulin. It is also found as a trace element in vegetables.

10. Vitamins. Vitamins are substances needed in small amounts for the normal functioning of the body. The name given to them by Sir Frederick Gowland Hopkins, the person who first realised their significance, was "accessory food factors".

(a) *Vitamin A*—needed for normal growth, and particularly for the growth and regeneration of epithelial tissues (tissues on the surfaces of organs). A deficiency leads to night-blindness (the inability to see in dim light) and to a decrease in the resistance to infection.

It is fat-soluble, so is found in fish-liver oils, butter, milk, eggs, and animal fats. Its precursor is carotene, one of the red pigments found in plants, so that carrots and tomatoes are also valuable sources.

(b) *Vitamin B₁* (*thiamin*)—required for the proper metabolism of carbohydrates. A deficiency causes the disease *beri-beri*, which is associated with eating polished cereals (especially rice) as the staple diet. The disease is characterised by nervous disorders and possibly heart-failure.

This vitamin is water-soluble. Its main sources are yeast, and rice and wheat germ (consequently wholemeal bread is a good source). It is also found in eggs, liver, and pulses.

(c) *Vitamin B₂ complex*—contains two main constituents, riboflavin and niacin, both of which are needed for the proper functioning of the oxidising enzymes. A deficiency of *riboflavin* first shows itself as cracking of the skin around the mouth and an inflammation of the cornea. Niacin (*nicotinic acid*) also forms part of the vitamin B₂ complex, the deficiency of which causes *pellagra*, a nervous disorder accompanied by chronic diarrhoea. The human body cannot convert nicotine inhaled from tobacco smoke into niacin, so that smoking is not a source of the vitamin.

The vitamin B₂ complex is also water-soluble. Yeast, meat, liver, milk, eggs and green vegetables are good sources. Maize (corn) is a particularly poor source.

(d) *Vitamin C* (*ascorbic acid*)—needed to keep the walls of the blood vessels healthy and increases the resistance to infections, *e.g.* to colds and influenza. A deficiency (scurvy) leads to bleeding from the gums and internal organs. The disease is prevented easily, by eating fresh fruit and vegetables. The vitamin is readily oxidised, so that cutting and boiling vegetables greatly reduces their vitamin C content.

Vitamin C is water-soluble. It is found in green vegetables, *e.g.* cabbage, citrus fruits, tomatoes, black currants, and rose hips. Potatoes are one of the chief sources in temperate countries.

(*e*) *Vitamin D*—a complex of vitamins needed for the absorption of calcium and phosphorus from the intestine, and consequently for the healthy development of bones and teeth. The malformation caused by its deficiency is called rickets (*see above* calcium). The vitamin can be manufactured in the skin in the presence of sunlight (ultra-violet light). Rickets was common in slum areas where children were living in crowded conditions, deprived of foods containing the vitamin (because they were expensive) and of sunlight.

Vitamin D is fat-soluble. It is found in fish-liver oils, butter and eggs.

Vitamins A and D are added to margarine (which is made from plant oils) to make it comparable with butter.

(*f*) *Vitamin E* (*tocopherol*)—needed for normal reproduction. Its absence causes abortion by females and loss of fertility in males.

Vitamin E is fat-soluble. It is found in wheat germ oil, green vegetables, eggs and liver.

(*g*) *Vitamin K*—needed for the clotting of the blood. This vitamin is fat-soluble and is found plentifully in green vegetables.

THE BALANCED DIET

11. The need for a balanced diet. We have seen that one of the prime functions of food is to provide energy. Theoretically all the energy required could be provided by eating carbohydrate; but this would mean that no body-building elements, *i.e.* amino acids, were taken in, and a protein-deficiency would result. This, in fact, happens in large areas of the world, especially in West Africa and South-East Asia where animal protein is scarce and the staple diet is made up of grains, *e.g.* rice and millet. A disease called *kwashiorkor* results from protein-deficiency and is prevalent among young children in these areas.

A check of the previous list of vitamins will show that animal fats and meat are good sources of vitamins, so that a diet lacking in animal products is frequently deficient in vitamins. The disease *trachoma*, which results in blindness, is common in West Africa. This is due to lack of vitamin A, and is associated with a low-protein diet. This is not to say that one cannot be healthy living on a vegetarian diet, but it must include vegetable sources of all the vitamins.

At the other end of the scale, the diet of the Eskimoes is almost entirely protein (seal meat and fish), and they rarely suffer from vitamin deficiencies.

In considering the structure of a balanced diet, the palatability and appearance of the food must be considered—a handful of starch and protein flakes with a few vitamin tablets would make a most uninteresting meal, and people would not eat it. Local customs also play an important part in deciding the structure of the diet. In some parts of the world good sources of vitamins are ignored as foods because they are taboo through religious or social custom. Similarly, the increasing amount of fried potatoes and sweets in the diet in the United Kingdom is largely responsible for the increase in the number of overweight children. Education can play a large part in persuading people to eat the foods which will do them most good.

12. Dietary requirements.

An adult man with a sedentary occupation, *e.g.* an office worker, needs about 3,000 Cal/day, while a labourer, doing heavy manual work, may need 5,000 Cal/day.

A housewife needs about 2,600 Cal/day.

A child of 3 years old needs about 1,400 Cal/day, but by the time he is 12–13 years old, he will need as much energy as his parents.

As a person becomes older, tissues are not renewed so quickly, they are less active, and their metabolic rate drops. They therefore need less food. Smaller meals are required, but they should still be nutritious. It is one of the dangers to old people, particularly when they are living on their own, that they "cannot be bothered" to prepare proper meals for themselves, and malnutrition results. Sometimes old people cannot afford the highly-priced protein-containing foods (*e.g.* meats) they require, and so make do with the poorer quality foods. This also causes malnutrition.

During pregnancy a woman's nutritional requirements

increase. She needs to ensure that she is eating sufficient proteins, minerals and vitamins for herself and the developing baby. After the birth of the baby, the production of milk is also a drain on the mother's food reserves, so that while she is feeding the baby a similar diet is essential.

Much of the energy produced from the food is used in maintaining the body temperature, consequently more energy-producing foods are required in a cold climate than a hot one. The necessary adjustment is best made by varying the carbohydrate intake.

Over-eating can be as harmful as malnutrition, as it leads to obesity. This is a result of excess carbohydrate being stored as fat. The excess weight which has to be carried can place a strain on the heart, which leads to heart failure. The fat which is stored around the heart reduces its efficiency with the same effect. Over-eating is often associated with the intake of food rich in animal fats, *e.g.* butter. This increases the cholesterol content of the tissues which may be a contributory factor to heart diseases.

The average adult man (3,000 Cal/day) needs about 500 g of carbohydrate, 85 g of fat, and 113 g of protein a day.

Lean meat contains about 20 per cent fat, 15 per cent protein; wheat bread—50 per cent carbohydrate, 1·5 per cent fat, 9 per cent protein; Eggs—10 per cent fat, 12 per cent protein; green vegetables (*e.g.* cabbage)—5 per cent carbohydrate, 1 per cent protein, 0·5 per cent fat; soya beans— 38 per cent protein, 18 per cent fat, 29 per cent carbohydrate; potatoes—20 per cent carbohydrate, 2 per cent protein; milk— 5 per cent carbohydrate, 4 per cent fat, 3·5 per cent protein.

The remaining percentages are small amounts of minerals, and mostly water.

THE DIGESTIVE SYSTEM

13. Introduction. The food of any animal is made up of large molecules which have to be broken down into smaller ones so that they can be absorbed into the body, transported to the cells which use them, and absorbed across the cell membranes. The processes involved in absorbing the food into the body take place in the digestive system. These processes can be divided into four stages:

(a) *Assimilation*—the taking-in of food into the digestive system, and breaking it down mechanically.

(b) *Digestion*—the enzymatic break-down of large molecules to smaller ones by hydrolysis.

(c) *Absorption*—the carrying of the small molecules across the gut wall into the blood system, so that they can be transported to the cells.

(d) *Defaecation*—the removal of the undigested parts of the food from the body.

14. Assimilation. Food is taken into the body through the *mouth* (*see* Fig. 11) into the *buccal cavity*. It first passes

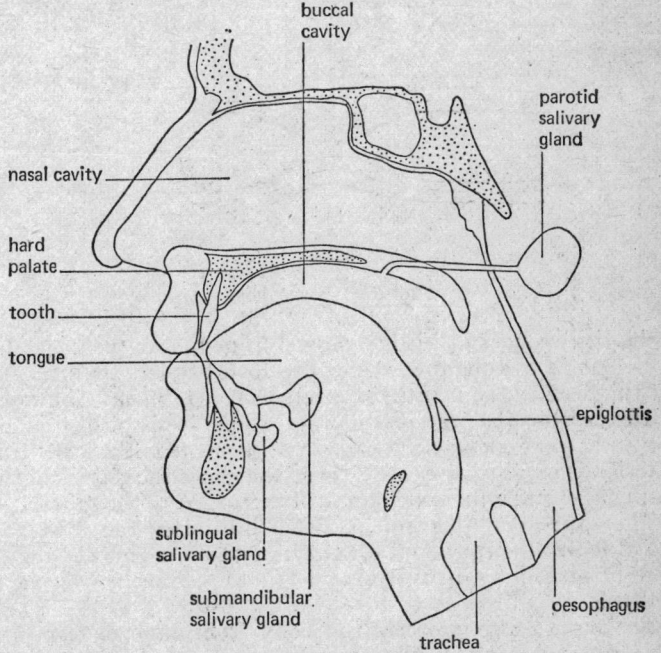

FIG. 11. Section through head.

through the *lips*, which are soft and muscular, and play an
important part in pushing the food through the mouth. Once
inside the mouth the food is masticated. During this process
it is broken down into smaller particles and lubricated by the
addition of *saliva*. The *teeth* and *tongue* play an important part
in mastication, the teeth in chewing the food, and the tongue
in squashing it against the *hard palate*. After mastication, the
soft ball of food (*bolus*) is swallowed. This in itself is quite a
complicated process. A whole series of muscles are involved,
including the tongue, and as the process is initiated, the
epiglottis closes over the top of the *trachea* to prevent the food
passing into the air passages. If food does accidentally "go

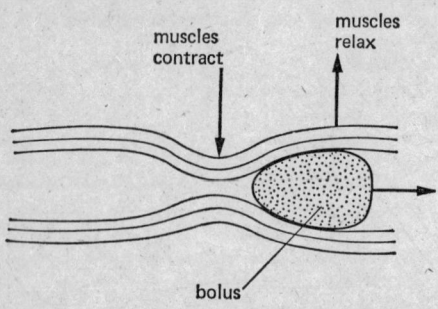

FIG. 12. Peristalsis.

down the wrong way" it is removed from the air passages by
a violent fit of coughing. Once the food has left the *pharnyx*
(at the back of the mouth) it enters the *oesophagus*. The walls
of the oesophagus are muscular and the bolus is pushed along
it by waves of contraction called *peristalsis* (*see* Fig. 12). This
reaction is involuntary (*see* V, **7**) so that once the food is in the
oesophagus, it cannot be voluntarily returned to the mouth.

The *saliva* contains mucin which lubricates the food and
salivary amylase (ptyalin), an enzyme which begins the digestion of starch by hydrolysing it to the disaccharide sugar
maltose.

The mouth also protects the body from damage from the
food, *e.g.* poisoning or extremes of temperature. Before entering the mouth, the food is tested by the lips for temperature

and consistency. When the food is in the buccal cavity it is tested by the tongue by tasting, and if it is unpalatable it is spat out. Tasting in the wider sense is two processes involving the tongue and the organs of smell (*olfactory organs*). This is why you cannot "taste" a food when you have a cold. The nasal passages are blocked with catarrh, and the odours from the food cannot reach the organs of smell. (The organs of taste will be discussed in the section on sense organs.)

15. Teeth. The teeth develop from small groups of cells below the epidermis which covers the jaw bones. As the teeth develop, they grow down into the sockets already developed in the jaw, so that the mature teeth are embedded in a socket in the jaw bone. During their development the teeth also grow upwards to cut their way through the gums (*see* Fig. 13).

Human beings produce two sets of teeth during their life, the *primary* or *milk teeth*, and the *permanent teeth*. The primary

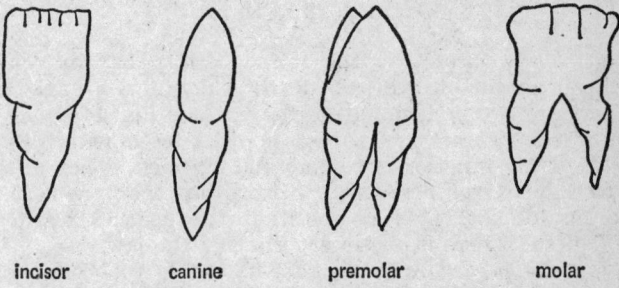

incisor   canine   premolar   molar

FIG. 13. Types of teeth.

incisors (front teeth) appear first when a baby is about 6 months old, and the teeth continue to erupt successively until the second molars appear at 2 years old. The primary teeth remain until the child is 6–7 years old when they begin to be displaced by the permanent teeth. The replacement begins with the incisors and continues towards the back of the mouth until the last primary molars (*premolars*) are replaced by the time that the child is 12 years old. As the jaws are enlarging during growth, areas develop behind the premolars where no primary teeth have developed; here new teeth (*molars*) are formed. The first molars appear at 6–7 years old, the second molars at 11–13 years, while the third molars (wisdom teeth) do not emerge until 17–22 years, or even much later.

There are twenty primary teeth and thirty-two permanent teeth in man. These are usually represented by a *dental formula*. The dental formula is written to represent one-half of the head. The teeth are represented by the letters I (incisors), C (canines), P or Pm (premolars), M (molars). The top figure in the following fraction is the number of teeth in the upper half-jaw, and the bottom figure the number in the lower half-jaw. Thus, the dental formula for primary teeth is

$$\text{I}\frac{2}{2},\ \text{C}\frac{1}{1},\ \text{P}\frac{2}{2} \quad \text{10 teeth in one half of the head} = 20 \text{ in}$$

total, and for the permanent teeth is

$$\text{I}\frac{2}{2},\ \text{C}\frac{1}{1},\ \text{P}\frac{2}{2},\ \text{M}\frac{3}{3}, \ i.e.\ 16 = 32 \text{ teeth in all.}$$

(*a*) *The incisors* are the front teeth. They are chisel-shaped for cutting pieces of food. They have a single root. There are two in each half of the jaw (*see* Fig. 14).

(*b*) *The canines* are strong, pointed teeth with a single root. Their function is to grip and rip food. They are not particularly well developed in man, but they are in carnivorous animals. This is associated with the need to grip and tear flesh. There is one in each half of the jaw.

(*c*) *The premolars* (*bicuspids*) have two roots. There are two in each half-jaw, lying behind the canines. The surface of the tooth is thrown up into sharpened ridges (cusps) which are used to grind and cut the food when the lower jaw is moved from side to side.

(*d*) *The molars* (*tricuspids*) lie behind the premolars. Those in the upper jaw have three roots. Their surfaces are cusped, but their cusps are less sharp than those of the premolars. These teeth are used for grinding the food.

The part of the tooth which is seen in the mouth is the *crown*, and the part embedded in the jaw is the *root*. The surface of the tooth is covered by a very hard *enamel* which is secreted by the *dentine* beneath. The dentine is much softer

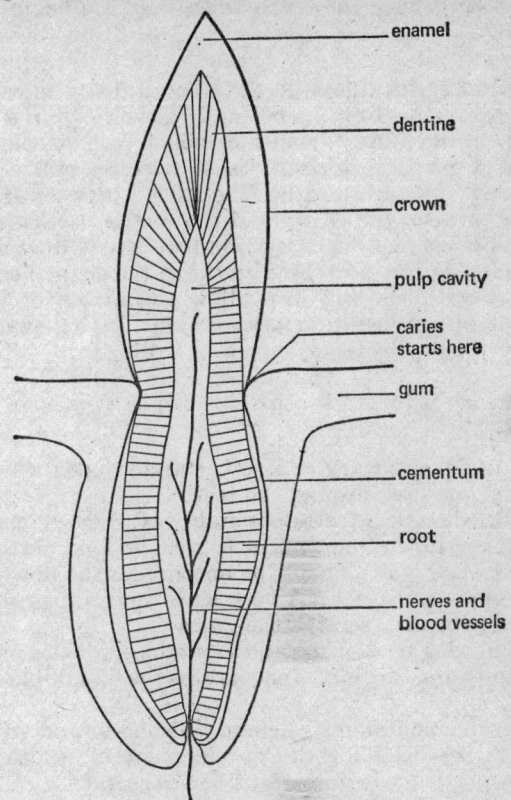

FIG. 14. Section through incisor.

than the enamel and is porous. This is the living part of the actual tooth and is the part subject to decay. In the centre of the tooth is the soft *pulp* which is made up of blood vessels and nerves. The blood vessels provide the necessary food materials for the tooth. The root is held in the jaw socket by the *cementum*. This is made up of minute fibres, which suspend the tooth rather than glue it firmly. Thus the cementum acts as a shock absorber for the tooth when they are being used for chewing.

16. Dental caries (tooth decay). Tooth decay is caused by bacteria attacking the enamel and, ultimately, the dentine and pulp cavity. Any condition which will encourage the growth and multiplication of these bacteria will encourage tooth decay. Bacteria can become lodged between the teeth. The zone between the crown and the root is particularly susceptible because any bacteria lying here have direct access to the dentine (*see* Fig. 14). Fine cracks in the enamel caused by chewing hard foods will also allow the access of bacteria. Bacteria multiply best in sugary solutions, so that sweet foods, especially "chewy" sweets, should be avoided.

17. Care of the teeth. Tooth decay can be avoided or reduced by:

(*a*) Avoiding sugary and soft, starchy foods which leave residues deposited around the teeth.

(*b*) Brushing the teeth, preferably after every meal, with an "up and down" movement to remove food particles and massage the gums. Massaging encourages the flow of blood to the gums and teeth. A very hard tooth brush should be avoided as it may scratch the gums.

(*c*) Chewing fresh fruits and vegetables, which exercise the teeth and gums. This also encourages the flow of blood to the teeth.

(*d*) A diet containing calcium, phosphorus and vitamin D, which is essential for the development of healthy teeth. This is particularly important for expectant mothers and young children. The expectant mother is providing calcium for her unborn child, and a deficiency in her diet may cause her teeth to suffer as her calcium reserves are passed on to her child.

18. Digestion. During digestion the carbohydrates are broken down to glucose and other monosaccharide sugars; the fats to fatty acids and glycerol; and the proteins to amino acids. These processes can be carried out outside the body, but then they need extreme conditions, *e.g.* we have already seen that to produce glucose from sucrose, it has to be *boiled* with hydrochloric acid. In the body such processes are carried out under much less extreme chemical conditions. This is possible because they are brought about by *enzymes*.

Enzymes are biological catalysts. They are capable of building-up or breaking down the molecules necessary for the tissues. They are necessary for *every biological process*, not only digestion. They have the effect of reducing the extremity of the conditions necessary for a particular reaction to take place, *e.g.* in the intestine sucrose is digested to glucose and

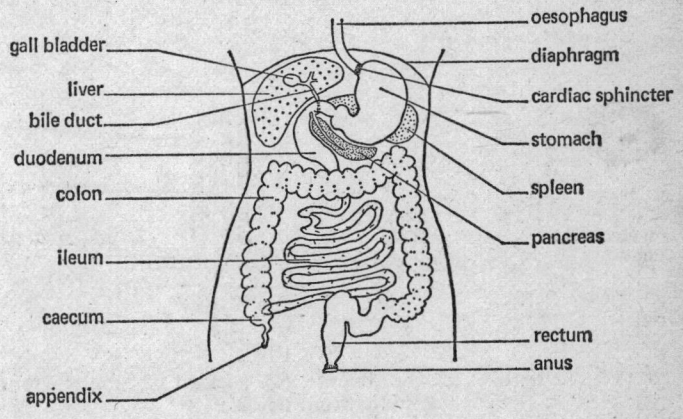

gall bladder — oesophagus
liver — diaphragm
bile duct — cardiac sphincter
duodenum — stomach
colon — spleen
ileum — pancreas
caecum — rectum
— anus
appendix

FIG. 15. Digestive system.

fructose at body temperature in mildly alkaline conditions. Enzymes are very *specific* in their action. One enzyme will only catalyse one reaction, *e.g.* an enzyme which will break down sucrose will not break down maltose. Enzymes are themselves proteins, so that extremes of temperature, acidity and alkalinity will destroy them (just as boiling solidifies the "white" of an egg).

The internal walls of the *stomach* and *small intestine* are thrown up into *villi* (*see* Figs. 15, 16). These are microscopic, finger-like processes which give the inner surfaces of these organs a structure rather like the pile on a carpet. The presence of villi greatly increases the surface area of the organs so that the digestion and absorption of food can proceed more quickly. If there were no villi, the intestine would have to be very much longer than it is, or the whole body metabolism would have to be different. As it is, the gut of an average man is about 10 metres long.

The outer wall of the intestine is made up of longitudinal and circular muscles, which keep the food moving through the intestine by peristalsis.

Food passes down the oesophagus which terminates at the *cardiac sphincter*. This is a ring of muscles which closes off

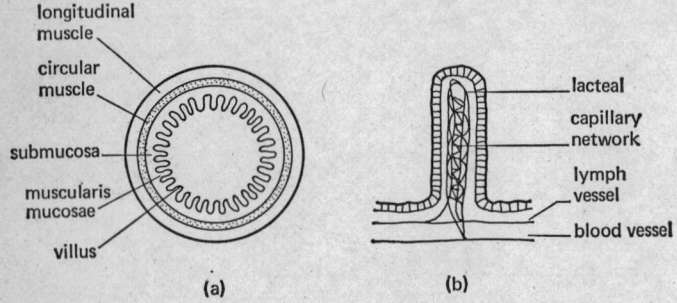

Fig. 16. Structure of the ileum: (*a*) section through ileum; (*b*) villus from ileum.

the top end of the stomach. The sphincter opens and the food passes into the stomach.

The *stomach* is a muscular sac, which, when empty, is about the size of two fists put together. The motions of the muscular walls *churn* the food, making it into a semi-liquid *chyme*. Glands in the stomach wall secrete hydrochloric acid, rennin, pepsin, gastric lipase, and gastrin.

Hydrochloric acid makes the food (which is slightly alkaline by the addition of saliva) acid. The acidity is necessary for the action of the stomach enzymes, and has some effect in destroying harmful bacteria.

Rennin is an enzyme which coagulates milk protein, and so is particularly important in the stomach of a young child.

Pepsin breaks down *proteins* to *peptones*. Peptones are molecules, not as large as proteins, but still containing a large number of amino acids.

Gastric lipase begins the digestion of fats to fatty acids and glycerol.

Gastrin is a hormone which stimulates the glands of the stomach to produce their enzymes.

Water is also secreted, which adds to the fluidity of the chyme.

Alcohol is absorbed directly by the stomach wall. This is why it is not advisable to take alcohol on an empty stomach.

The chyme leaves the stomach through the *pyloric sphincter* and passes to the duodenum.

The *duodenum* is about 30 centimetres (cm) long, and within its loop lies the *pancreas*, the duct of which opens into the duodenum. The *bile duct* also opens into the duodenum.

The pancreas produces the following enzymes:

(a) *Amylase*, which digests starch to maltose.
(b) *Lipase*, which digests fats to *fatty acids* and *glycerol*.
(c) *Trypsinogen*, a precursor of trypsin.

The hormone *insulin* is also produced by the pancreas in the *Islets of Langerhans*. (This will be discussed in detail in the section on hormones—*see* V, **43**.)

The walls of the duodenum produce a digestive juice (*succus entericus*). This contains:

(a) *Enterokinase*, which activates trypsinogen to the enzyme trypsin. This enzyme digests proteins to *amino acids*.

(b) *Maltase*, *sucrase*, and *lactase*, which digest the appropriate disaccharide sugars to *monosaccharides*.

Thus, by the time that the food leaves the duodenum it is completely digested to fatty acids, glycerol, monosaccharides and amino acids, ready to be absorbed.

Bile is produced in the liver, and contains the breakdown products of worn-out red blood cells. It is stored in the *gall bladder*. Its function in the intestine is to make the food alkaline and to break down (*emulsify*) large fat droplets to smaller ones.

19. Absorption. Absorption takes place in the *ileum*. This is about 5 m long with the inner surface thrown up into villi. Within each villus are very fine blood vessels and lymph vessels (*lacteals*). The digested foods diffuse or are actively transported by the "pumping action" of their cells across the fine walls of the villi. The amino acids and monosaccharides enter the blood vessels while the fatty acids and glycerol enter the lacteals. Within the lacteals the fatty acids and glycerol re-combine into fat *droplets*.

The blood vessels in the villi join to form a fine network of larger vessels which run through the mesentery between the loops of the intestine. These blood vessels ultimately join to form the *hepatic portal vein* which carries blood from the intestine to the liver. (The function of the liver is discussed in a later section—(*see* VI, **10**).)

From the villi branches of the lacteals fuse to form a lymph vessel which joins the vena cava just before it enters the heart.

Having the most of the digested food removed from it, the intestinal contents pass into the *large intestine*. At the junction of the small intestine and the large intestine is the sac-like *caecum* with its attached *appendix*. These organs have no obvious significance in man, but in herbivorous animals they are relatively much longer, and contain bacteria which digest cellulose. From the caecum the food passes into the large and wide *colon* which runs upwards as far as the diaphragm, along the right side of the body and descends along the left side to terminate in the short muscular *rectum*.

Water is absorbed from the intestinal contents through the walls of the colon and into the blood vessels. This conserves water which would otherwise be lost, and dries the faeces, reducing their bulk.

20. Defaecation. The food, having been partly dried, passes into the *rectum* from which it is voided periodically through the *anus*. A ring of sphincter muscles control the opening of the anus.

NOTE: Defaecation is not strictly speaking a form of excretion (*see* VI, **13**), because the faeces are the unused part of the food originally taken in through the mouth. They have never entered into the metabolism of the body. Some substances, *e.g.* bile and water, introduced into the food during its passage through the intestine are excretory products, and these are removed incidentally with the faeces.

Difficulty in defaecating is called *constipation*. The direct cause of this is a sluggish contraction of the muscles of the rectum and large intestine. One of the common causes of this complaint is the lack of undigestible material (bulk) in the food, so that a diet containing "roughage", *e.g.* fresh fruit and vegetables, may cure the complaint. Other causes of the complaint include irregular times of meals and nervous tension. Serious cases of constipation can be treated with purgatives. These substances activate the muscles of the large intestine and rectum, sometimes violently. Mild purgatives (*e.g.* liquid paraffin) are called *laxatives* while the more severe ones (*e.g.* castor oil, senna, and phenolphthalene) are called *apperients*. Enemas (irrigating the lower intestine with water through the rectum) are also used to treat constipation.

When the faeces are discharged before the normal amount of water is removed the condition is called *diarrhoea*. This may be caused by eating too much fruit etc. containing purgatives, *e.g.* figs, prunes, or rhubarb. Diarrhoea may be the symptoms of some other complaint, *e.g.* a chill, or some more serious disease like amoebic dysentery, bacterial dysentery, typhoid fever, or cholera.

21. Some experiments to show the conditions necessary for enzyme action.

(*a*) To show the conditions necessary for the reaction of salivary amylase (ptyalin).

Put about 3 cm^3 of saliva into three test-tubes and an equal quantity of water into a fourth as a control (D). Leave one tube (A), boil the second tube (B) and add a few drops of hydrochloric acid to a third (C). Add equal quantities of starch solution to each tube. Shake and leave for 15 min. at room temperature. Divide the contents of each tube into two, and test one of each pair with iodine solution, and the other with Fehling's solution (or Benedict's solution). Tube A will give a lighter starch reaction and a positive Fehling's reaction. The other three test-tubes will give a negative Fehling's reaction and a deep blue-black starch reaction.

(b) To show the conditions necessary for the reaction of pepsin.

Place about 3 cm³ of pepsin solution in each of three test-tubes, and an equal quantity of water in a fourth as a control (D). Leave one tube (A), boil the second tube (B) and add a few drops of hydrochloric acid to the third (C). Add a little egg white (protein) to each tube and leave to stand for as long as possible. Test each tube with Millon's reagent. Only tube C will give a negative reaction, showing that the enzyme is destroyed by boiling, and that only under acid conditions will the enzyme digest protein.

RESPIRATION

22. Introduction. Respiration is the process whereby energy is released in the living cell. It is a chemical process whereby the energy stored in large molecules is released for use in cells. This chemical process is sometimes called *tissue* or *cellular* respiration to distinguish it from *breathing* or *external* respiration. (The process of breathing will be considered in VI.)

23. Aerobic respiration. The respiration of a hexose sugar is represented by the over-all equation:

$$\underset{\text{sugar}}{C_6H_{12}O_6} + \underset{\text{oxygen}}{6O_2} \rightarrow \underset{\substack{\text{carbon}\\\text{dioxide}}}{6CO_2} + \underset{\text{water}}{6H_2O} + \text{energy}$$

You will see that for each molecule of sugar, six molecules of oxygen are required for the release of energy.

The same equation would apply if the sugar was burnt and the energy dissipated as heat. This would be useless in living organisms, so here the process proceeds in steps, and the energy released is captured by the formation of chemical bonds.

There are two main stages in aerobic respiration: (a) the formation of pyruvic acid, and (b) the breakdown of the pyruvic acid in the Kreb's cycle. The first stage does not need oxygen, but the second stage does. During both these stages carbon dioxide is released from the sugar. The oxygen is required to oxidise the hydrogen from the sugar to water.

The energy released is used to convert adenosine diphosphate (ADP) to adenosine triphosphate (ATP) by forming a new chemical bond between two phosphate groups:

adenosine –P–P + P + energy $\longrightarrow$ adenosine–P–P–P

(P is used to represent a phosphate radical.)

The energy held in the ATP is released into the cells re-forming ADP which can be used again to 'hold' more energy.

24. Anaerobic respiration.
It is possible for tissues to live for a short time without oxygen. (The brain is most easily affected by oxygen shortage. Permanent damage to the brain can result from only a few minutes' absence of oxygen.) Under these conditions anaerobic respiration takes place. The end-product of anaerobic respiration in human tissue is lactic acid, and less energy is released from the sugar than during aerobic respiration.

sugar $\longrightarrow$ lactic acid + carbon dioxide

Anaerobic respiration takes place in overworked muscle, *e.g.* when running or doing some extreme form of exercise, as the oxygen supply is not sufficient for the complete oxidation of the sugar. The lactic acid accumulating in the muscle results in *fatigue*, and the muscles are prevented from working efficiently. At the end of the exercise, this lactic acid has to be removed rapidly, needing an excessive supply of oxygen to make up for the previous oxygen lack (the *oxygen deficit*). This results in panting. The rapid release of now unwanted energy causes the body to get hot, resulting in sweating and flushing of the skin.

THE HYGIENIC PREPARATION AND STORAGE OF FOOD

25. Introduction.
In days gone by most foods were produced and consumed locally. There were few methods of preserving food so that life was a succession of glut and famine. Such a situation exists in many countries today. Nowadays foods can be preserved by many methods, which means that there can

be an even distribution throughout the year, and that foods can be transported over great distances.

Foods become unfit for human consumption because:

(a) *they become contaminated* by bacteria and other micro-organisms;

(b) *they decay* through the liberation of enzymes within their cells (*lysis* of the cells);

(c) *fats in them become oxidised* and so they become rancid.

The methods employed in preserving food prevent or retard these happenings.

26. Methods of preserving food.

Whatever method of preservation is employed, the food must retain its nutritive value and it must be palatable.

(a) *Salting.* This is a very old method of preserving foods, especially meats. The high concentration of salt in the food dehydrates any micro-organisms present.

(b) *Pickling.* This is done by soaking the food in brine or vinegar which produces conditions in which the micro-organisms cannot survive. The flavour of the food is altered, but it is still palatable.

(c) *Smoking.* Fish, especially herrings, are smoked, *e.g.* kippers and bloaters. Smoking dehydrates the food. Particular woods, *e.g.* oak, are used for smoking to give a characteristic flavour to the food.

(d) *Bottling and canning.* The principle in both these methods is the same. Cooked food is placed into sterilised containers and covered with boiling water or syrup. The vapour from the liquid fills the space above the liquid and the top is sealed on. When the vapour cools, it condenses, leaving an air-free space above the food. The heating kills the micro-organisms and, as the air is driven out of the container, no more come in contact with the food. Cans are lacquered on the inside to prevent acids in the food from attacking the metal. Foods preserved in these ways lose some of their vitamins, and their flavours are markedly different from when they are fresh.

(e) *Dehydrating.* Micro-organisms need water to multiply,

so that the removal of water from the food greatly retards
the spoilage of food. Sun-drying is a very old method of
preserving food, particularly in warm countries, *e.g.* meat,
raisins, dates, grain.

Artificial drying has not been very successful in the past
as the flavour of the food is lost. The introduction of freeze-
drying and vacuum-drying has enabled the food to be
dehydrated without heating, so that the flavour can be
retained. The production of dried milk and eggs during the
Second World War lent impetus to discovering new methods
of drying foods which has resulted in the production of
dried soups, "instant" potato, "instant" coffee, etc.

(*f*) *Refrigeration.* Cooling retards the multiplication of
micro-organisms. Frozen foods retain their colour and
flavour better than foods preserved in other ways. Deep-
frozen foods (kept at $-10°C$) keep longer than those kept
in a refrigerator because the growth of micro-organisms is
completely retarded. The difficulty in deep-freezing food
lies in the necessity for rapid freezing. Slow cooling causes
the formation of large ice crystals in the cells of the food
(particularly fruit and vegetables). These break the cells so
that the food's texture is destroyed.

Once the food is thawed it should be eaten quickly, and
not re-frozen.

(*g*) *Pasteurisation.* This method, discovered by Louis
Pasteur, is used to kill bacteria in liquids (especially milk)
the flavour of which would be destroyed by boiling. There
are two methods:

(*i*) The liquid is heated to 65°C and kept at this temperature for 30 min. before being cooled rapidly to 10°C;

(*ii*) the liquid is heated to 71·5°C and kept at that temperature for 15 sec. before being rapidly cooled to 10°C.

27. Food hygiene.
The points mentioned below reduce the
risk of food becoming contaminated by disease organisms.

(*a*) Hands should be washed with soap and water before
touching food, especially after using the toilet.

(*b*) Wounds, boils, etc. should be covered with clean
dressings.

(*c*) Coughing and sneezing over food should be avoided.

(d) Clean overalls and caps should be worn.

(e) Displayed foods should be covered.

(f) Pets should never be allowed near food.

(g) All food should be stored in a cool, dry place, free from vermin and flies.

(h) Surfaces on which food stands should be washed frequently and should have no cracks in them which could harbour bacteria.

(i) Cracked crockery and cooking utensils should be discarded.

(j) Refuse bins should be properly covered, dry, and raised off the ground. This prevents access by rats and flies.

28. Legal enforcement. Most of the points mentioned above are legally enforceable in places where food is sold for consumption under the Food and Drugs Act 1955. Government inspectors regularly check restaurants and other places where foods are kept and served. Cases of food poisoning should be notified to the local Health Authorities so that the source of infection can be traced. The Food and Drugs Act also controls the various additives, e.g. colouring and artificial flavourings which may be added to foods.

29. Milk. Milk is an exceptionally good medium for the multiplication of bacteria and so is a good medium for the spread of diseases. The causal organisms of a wide variety of diseases and the various forms of food poisoning may be carried on the hands and clothing of an infected person. Tuberculosis, which infects cows, can be transmitted through the milk to human beings. The disease has been greatly reduced in people in the U.K. since the introduction of the tuberculin-testing (T.T.) scheme in 1938. Herds are tested for tuberculosis, and no new cattle are allowed to be introduced to the herds unless they have also passed the test.

To reduce the chances of milk becoming infected, rigorous legislation is in force in many countries controlling the conditions under which milk may be handled. This ensures the cleanliness of the cows during milking, the cleanliness of the dairymen, and the cleanliness of the containers in which the milk is transported and sold.

30. Food poisoning. Apart from being the agent for the transmission of disease organisms, certain bacteria which live in food can cause food poisoning.

(a) *Clostridium botulinum*. This is a soil bacterium, but if it invades man it produces a toxin which causes paralysis of the nerves (botulism). Although this disease is rare it can be fatal. It can be avoided by making sure that foods, especially vegetables, are thoroughly washed.

(b) *Salmonella* spp. These bacteria live particularly on meat and eggs. Pre-cooked meat and ducks' eggs are particularly dangerous sources. The symptoms of this form of food poisoning are sickness and diarrhoea, both of which can be severe.

PROGRESS TEST 2

1. What is the process by which green plants produce their food? **(1)**
2. What must animals do to complex foods before they can absorb them? **(1)**
3. Draw a diagram of the carbon cycle. **(1)**
4. Give an example of a food chain. **(2)**
5. What is a calorie? How is this term used when referred to foods? **(3)**
6. What is the general formula for carbohydrates? **(5)**
7. What is the chief importance of carbohydrates in the diet? **(5)**
8. Name five foods which contain a high proportion of carbohydrate? **(5)**
9. Outline a test for starch. **(5)**
10. Outline a test for reducing sugars. **(5)**
11. Why are fats important in the diet? **(6)**
12. Name five foods which contain a high proportion of fats or oils. **(6)**
13. Name the elements contained in proteins. **(7)**
14. Define a first-class protein. **(7)**
15. Why are proteins necessary in the diet? **(7)**
16. Name five foods which contain a high proportion of proteins. **(7)**
17. What is a test for proteins? **(7)**
18. Give three reasons why water is essential in the diet. **(8)**
19. Why is calcium needed in the diet? **(9)**
20. Name three good sources of calcium. **(9)**
21. Why is phosphorus needed in the diet? **(9)**

22. State three good sources of phosphorus. **(9)**

23. Why is iodine needed in the diet? **(9)**

24. Why is iron needed in the diet? **(9)**

25. Name the fat-soluble vitamins. **(10)**

26. Name three good sources of each of the following vitamins: (a) A, (b) B₁, and (c) D. **(10)**

27. Why is vitamin A needed in the diet? **(10)**

28. Why is vitamin B₁ needed in the diet? **(10)**

29. Why is vitamin D needed in the diet? **(10)**

30. What are the calorie requirements of (a) a housewife, (b) a manual labourer? **(12)**

31. Name one harmful effect of over-eating. **(12)**

32. From the following select the best sources of (a) protein, (b) carbohydrate, and (c) fat.

       Lean meat; potatoes; wheat flour; soya beans; eggs. **(12)**

33. State three functions of the mouth. **(14)**

34. Draw a diagram of a tooth in longitudinal section. **(15)**

35. State three ways of avoiding tooth decay. **(16, 17)**

36. Define the term "enzyme". **(18)**

37. Name an enzyme which digests starch. **(14)**

38. Name an enzyme which digests proteins. **(18)**

39. Name an enzyme which digests fats. **(18)**

40. State the functions of the stomach. **(18)**

41. State the functions of the duodenum. **(18)**

42. State the enzymes produced in the pancreas. **(18)**

43. State the functions of the ileum. **(19)**

44. What are the functions of the large intestine? **(19)**

45. What is the cause of constipation? **(20)**

46. State the formula for aerobic respiration. **(23)**

47. What is the function of ATP? **(23)**

48. When does anaerobic respiration take place? **(24)**

49. Give three reasons why food becomes unfit for human consumption. **(25)**

50. State seven ways of preserving food. **(26)**

51. Why is food hygiene important? **(27)**

52. Name the Act which controls the sale of foods for consumption. **(28)**

53. Name three ways used to reduce the chances of milk becoming contaminated. **(29)**

54. Name two organisms which cause food poisoning. **(30)**

EXAMINATION QUESTIONS

1. Discuss why a balanced diet is necessary to good health.

2. Outline the part played by bacteria in: (a) tooth decay;

(b) food poisoning; (c) food contamination. Give two ways in which tooth decay can be avoided.

3. Describe an experiment to demonstrate the functioning of one named enzyme. Name three other enzymes, where they are produced, and what their functions are.

4. If you were a Public Health Inspector, what conditions would you insist on in a shop that was selling food?

5. What are the importance of the following in the diet: (a) vitamin A; (b) vitamin C; (c) vitamin D; (d) iron; (e) calcium. Name two good sources of each of the above.

CHAPTER III

THE TRANSPORT SYSTEM

1. Introduction. Small, unicellular organisms are in direct contact with their external environment, so that oxygen can diffuse into the cells directly. In larger organisms the majority of the cells are not in contact with the external environment, so that oxygen has to be brought to them through other tissues. Because the organisms are too large, oxygen cannot diffuse quickly enough from cell to cell so that some form of transport system is necessary. Similarly, digested food has to be moved from the site of digestion to the cells in which it is to be used. With the increase in size comes the specialisation of tissues and organs, some of which produce substances, *e.g.* hormones, which are effective at some site distant from that at which they are produced. These substances require transporting. The transport system of man is made up of the *vascular system* and the *lymphatic system*.

2. The vascular system. This system carries *blood*. It consists of a pumping *heart*, *arteries*, *veins* and *capillaries*.

The vascular system of a human being is a *closed, double* circulatory system. It is *closed* because the blood is confined to vessels and does not bathe the tissues directly, and it is *double* because the blood passes through the heart twice from the time it leaves an organ to the time it returns to it (*see* Fig. 17).

The blood is oxygenated in the lungs and passes through the *pulmonary veins* into the *left atrium* (auricle) of the heart. It passes downwards through the valves in the heart to the *left ventricle* from which it passes into the *aortic arch*. From the aortic arch branch the *carotid arteries*, carrying blood to the neck and head, and the *subclavian arteries* carrying blood to the arms. The aortic arch curves to run posteriorly along the back on the dorsal side of the body as the *dorsal aorta*. This gives off branches to the rib muscles (*intercostal arteries*), stomach (*gastric artery*), liver (*hepatic artery*), kidneys (*renal arteries*), gonads (*spermatic* or *ovarian arteries*), finally branching to form the two large *common iliac arteries* which supply the legs.

Blood is collected from the head and neck in the *jugular*

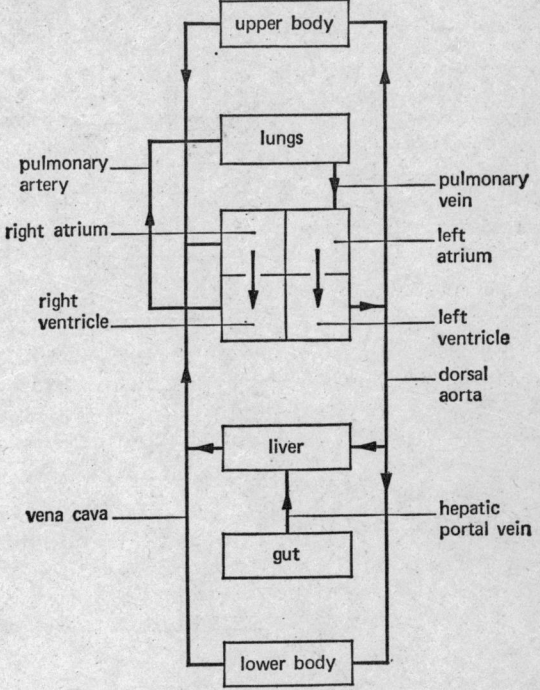

FIG. 17. Diagram of circulatory system.

veins and from the arms in the *subclavian veins*. These vessels converge into the *superior* (*anterior*) *vena cava* which runs into the *right atrium* (auricle) of the heart. Blood is collected from the other parts of the body into the *inferior* (*posterior*) *vena cava* which runs alongside the dorsal aorta and also enters the right atrium. The *iliac veins* drain the legs, the *spermatic* or *ovarian veins* the reproductive organs, the *renal veins* the kid-

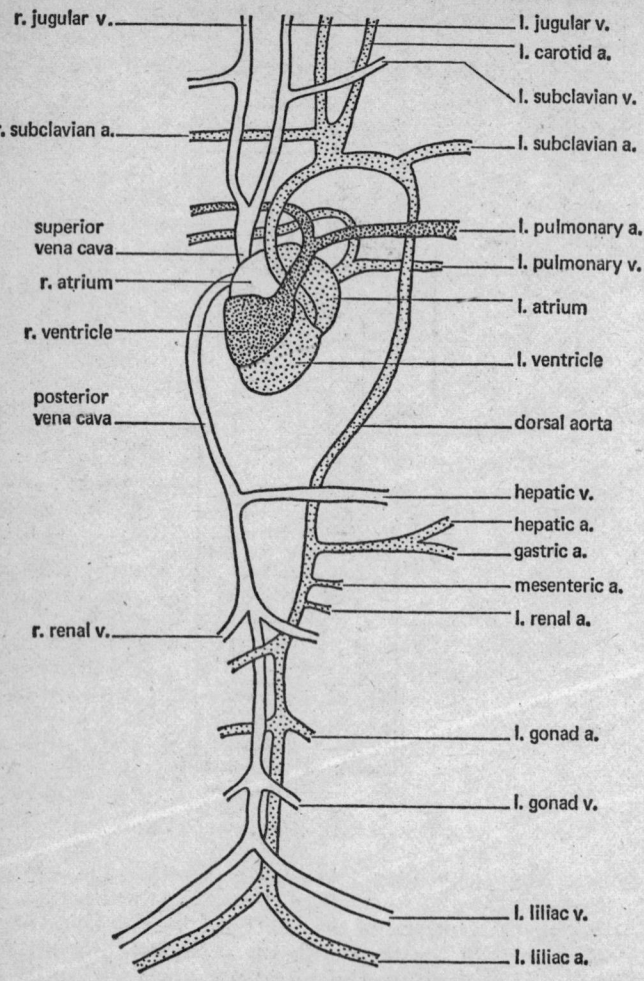

r. jugular v.
l. jugular v.
l. carotid a.
l. subclavian v.
r. subclavian a.
l. subclavian a.
superior vena cava
l. pulmonary a.
l. pulmonary v.
r. atrium
l. atrium
r. ventricle
l. ventricle
posterior vena cava
dorsal aorta
hepatic v.
hepatic a.
gastric a.
mesenteric a.
l. renal a.
r. renal v.
l. gonad a.
l. gonad v.
l. liliac v.
l. liliac a.

FIG. 18. Main blood vessels.

neys, the *hepatic vein* the liver, and the *intercostal veins* the rib muscles.

The intestine is drained by the *hepatic portal vein* which runs directly to the liver, so that food materials from the gut have to pass through the liver.

Blood from the right atrium is pumped through valves, into the *right ventricle* of the heart and from here it is pumped through the *pulmonary arteries* to the lungs (*see* Fig. 18).

The blood passes through very fine capillaries within the organs and it is through their walls that the interchange of materials between the blood and the cells of the tissues takes place.

THE HEART

3. Introduction. The heart is the pumping organ of the vascular system. It lies in the thoracic cavity on the central line of the body, with the lower point inclined to the left. Enclosing the heart is a tough membrane (*pericardium*) and from the heart arise the main blood vessels mentioned previously. In a dissection of a small mammal you will see the size of the heart relative to the lungs which surround it. Also notice that it lies high in the thoracic cavity, the lower end being beneath the fifth rib.

4. Structure. The heart is a very muscular organ. It is divided vertically by a wall of muscle. In each half there are two chambers, the *atria* (singular *atrium*) or *auricles* above, and the *ventricles* below.

The auricles are thin-walled, their muscles only having to pump the blood from the atrium to the ventricle. Ventricles have thick muscular walls. The pressure they generate has to be sufficient to force the blood away from the heart through the tissues. The wall of the left ventricle is much thicker than that of the right ventricle as it has to produce a pressure to force the blood throughout the body. The right ventricle has to force the blood only through the lungs (*see* Fig. 19).

Between the auricles and ventricles are tough membranous *valves* which prevent the backflow of blood from the ventricle to the auricle. The valve between the right atrium and ventricle is the *tricuspid valve* and between the left atrium and ventricle the *mitral valve*. Tough fibres (*chordae tendinae*) attach the valves to the walls of the ventricles to prevent the inversion of the valves into the atria when the ventricles contract. Further, *semi-lunar* (pocket) *valves* are situated at

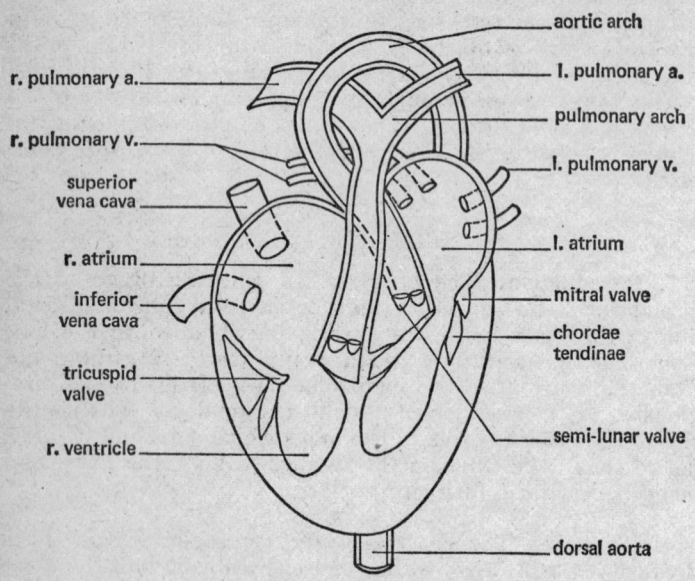

FIG. 19. Structure of the heart.

the base of the arteries. These prevent the backflow of blood from the arteries into the ventricles.

The heart is beating continuously from before a person is born until he dies. This continuous activity demands a special type of muscle (*cardiac muscle*). The individual cells can carry on beating indefinitely as long as they have a long rest period between each beat.

5. **Function.** The beating of the heart is co-ordinated by the contraction of its own muscles. The auricles contract in unison pumping blood into the ventricles, then the ventricles

contract pumping the blood into the arteries. The whole heart
then relaxes. The contractions are called *systoles* and the
relaxation is the *diastole*. The complete heart beat is atrial
systole, ventricular systole, diastole. It is these heart movements that are heard through a stethoscope. Any irregularity
in them can indicate a disease of the heart, or some other
medical condition. Each diastole lasts about half a second.
The adult's heart beats at a rate of about 75 beats a minute,
while that of a baby is beating at a rate of about 140 beats a

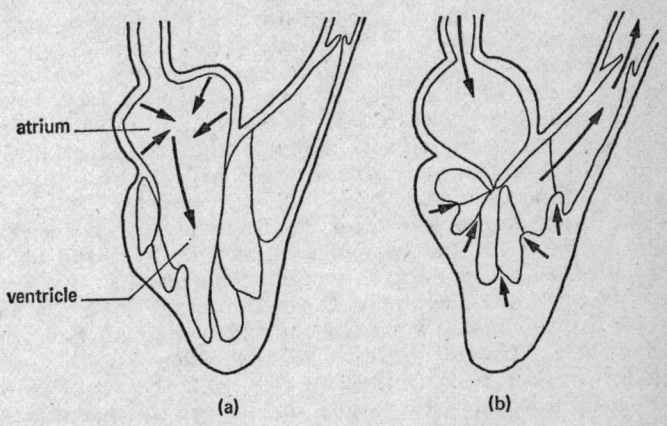

atrium

ventricle

(a)   (b)

FIG. 20. Beating of the heart: (a) systole of atrium; (b) systole of
ventricle.

minute. This rate can be measured by taking the pulse. In
terms of blood volume, at 75 beats a minute, the heart pumps
about 4 litres of blood through it each minute. During
vigorous exercise, the heart rate may be doubled and become
more vigorous, so that there is 15–20 l of blood passing through
it each minute (*see* Fig. 20).

For a fluid to move from one place to another, there must be
a difference in pressure between the two places. This is provided by the contraction of the ventricles. The blood pressure
in the main arteries, *e.g.* in the arm of a young adult, is about
880 mm of mercury. That is about 120 mm of mercury above

atmospheric pressure. It is lower in a child (about 860 mm of mercury) and higher in an older person. By the time a man is 65 years old, his blood pressure may be 910 mm of mercury. This increase in pressure is due to the thickening of the arterial walls and a loss in their elasticity. As the blood runs through the capillaries it meets with a resistance which reduces its pressure so that the venous blood pressure is much less than that of the arteries.

6. Heart diseases. Heart diseases account for a large number of deaths in countries with a high standard of living. Many people, particularly in middle age or later life, die of a "heart attack" or coronary thrombosis. A coronary thrombosis is a clot of blood blocking the coronary artery which provides the heart with blood. The formation of such blood clots has been associated with a high level of cholesterol in the blood, which in turn may be associated with a high intake of animal fats in the diet.

The thickening of the artery walls reduces their internal diameter so a higher pressure is needed to force the blood through them. This means a strain on the heart, so that sudden exercise or shock may cause the heart to overwork. Stress, continuous nervous tension and worry can have the same effect. Heart muscles, like any other muscles, need regular exercise to keep them in tone. The hearts of many people in sedentary occupations do not get this exercise, so that they cannot cope with any sudden exertion. Regular mild exercise, e.g. walking, keeps the heart muscles as well as the other muscles in tone.

Over-eating leads to the storage of fat around the heart, which inhibits its proper function and can lead to heart failure.

There is a connection between the incidence of heart attacks and the number of cigarettes smoked each day.

Regular exercise and moderate eating can reduce the chances of heart attacks, but there are other diseases which may be unavoidable, e.g. infection of the heart tissue or *pericardium*, and diseases affecting the valves.

Modern surgery is making great advances in the cure of heart diseases, e.g. heart transplants, artificial valves, and electronic "pace makers", while drugs are used to reduce blood pressure.

THE VESSELS

7. Arteries. These vessels carry blood away from the heart. The walls are made up of:

(a) An inner lining (*epithelium*);
(b) elastic fibrous tissues;
(c) muscles; and
(d) an external fibrous coat.

Thus the arteries are able to withstand the pressure exerted by the blood as it is pumped from the heart.

8. Veins. Veins carry blood to the heart. They have a wider bore than the arteries, and the walls do not contain elastic tissue. The pressure of the blood in them is relatively low.

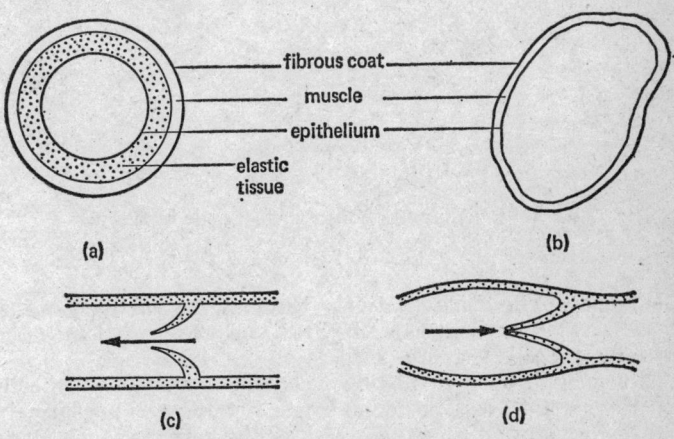

FIG. 21. Structure of blood vessels: (a) section of artery; (b) section of vein; (c) semi-lunar valve open; (d) semi-lunar valve closed.

The blood from the lower part of the body is flowing *upwards* through the veins to the heart, and the larger veins have *semi-lunar valves* in them to prevent the backflow of blood. The valves are obvious in the veins of the leg. Their collapse is called varicose veins (*see* Fig. 21).

9. Capillaries. As the arteries and veins penetrate the tissues they become finer until they form capillaries which join the arteries to the veins. The capillaries are exceedingly fine, with walls only one cell in thickness, and a bore such that only a single red-blood cell can pass along them at a time. The capillaries penetrate through the tissues so that every cell is in contact with a capillary. It is through the capillary walls and the walls of the cells that the exchange of dissolved substances takes place. The fluid from the blood, containing smaller molecules of dissolved substances, can pass through the capillary walls and bathe the tissues. This liquid is *lymph*. The lymph contains *lymphocytes* as well as the dissolved

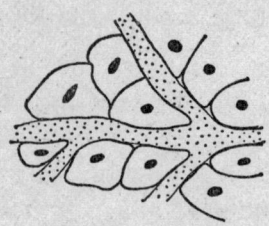

Fig. 22. Capillary in contact with cells in tissue.

substances. These pass actively through the capillary walls. Food, and other materials, *diffuse* from the lymph into the surrounding cells (*see* Fig. 22).

Oxygen and carbon dioxide also diffuse between the cells and the capillaries. The blood in the arteries has a relatively high concentration of oxygen, while the respiring cells have a relatively high concentration of carbon dioxide, so the oxygen passes into the cells, and the carbon dioxide passes out from the cells by diffusion.

Diffusion is the passage of dissolved substances or gases from a high concentration to a low concentration.

Capillaries can be seen in the web of a tadpole's foot. Anaesthetise a tadpole by adding a drop of chloroform to a dish of water in which it is placed. Look at the web of the foot under the microscope, and the blood will be seen moving through the capillaries.

THE BLOOD

10. Introduction. In this section we will discuss the blood as a transport system; its other functions will be mentioned in VII.

11. Functions. The following substances are transported by the blood:

(*a*) *Digested food.* Glucose and amino acids are dissolved in the plasma and transported from the small intestine to the liver, and hence to the rest of the body.

(*b*) *Oxygen* is transported by the red blood cells. In the relatively high oxygen concentration of the lungs, it combines with haemoglobin to form *oxyhaemoglobin*. When the blood passes through a respiring tissue, the oxyhaemoglobin breaks down to release the oxygen, and the haemoglobin is released to combine with more oxygen. The released oxygen diffuses in solution through the walls of the capillaries to the tissues.

(*c*) *Carbon dioxide*, produced by the respiring cells, diffuses into the capillaries. In the blood, it is transported partly combined with haemoglobin and partly in the plasma as a solution of sodium hydrogen carbonate. On reaching the lungs the carbon dioxide is released and diffuses into the air sacs.

(*d*) *Heat.* Respiration releases energy, some of which manifests itself as heat. This is particularly obvious in the case of active muscle, but much of the body's heat is generated during the metabolic activities of the liver. Heat produced in these ways is distributed around the body by the blood, and excess is dissipated through the skin (*see* VI).

(*e*) *Dissolved toxic metabolic by-products* are diffused from the cells into the plasma in which they are transported to the liver. In the liver they are made relatively harmless and discharged into the blood stream which carries them to the kidneys through which they pass as urine.

(*f*) *Hormones* pass from the glands which produce them

into the plasma, and are carried by it to the organs which they affect.

12. Diseases of the blood.

(a) *Anaemia*. There are two types of anaemia, both of which are characterised by general tiredness, pallor and lowering resistance to disease, and are due to the lowering of the number of red blood cells in the blood. Simple anaemia is common among fast-growing children and pregnant women. It is usually due to the inability of the bone marrow to produce enough red blood cells, or the lack of iron to manufacture haemoglobin. Pernicious anaemia affects old people and is due to the inability of the bone marrow to produce red blood cells to make up for those which are destroyed. Both these types of anaemia are treated by rest, iron tablets, and feeding or injecting with liver extract. In both these types of anaemia good, easily digestible food forms a valuable part of the treatment.

Anaemia is a symptom of other diseases, *e.g.* malaria, hookworm invasion, blood fluke. Lack of vitamin B_{12} or folic acid in the diet also are indicated by anaemia. Sickle-cell anaemia is a particular case which is inherited and is fatal.

(b) *Leukaemia*. This a fatal cancer of the blood. The cells producing the leucocytes become over-active so that there is an enormous increase in the number of white corpuscles in the blood.

THE LYMPHATIC SYSTEM

13. Structure. The lymphatic system consists of thin-walled vessels which penetrate the tissues in the same way as the capillaries. They join up to form two main ducts which join the vena cava at the junction of the jugular and sub-clavian veins. There are some semi-lunar valves in the larger lymph vessels to ensure a one-way flow towards the heart. In their passage to the heart, the lymph vessels pass through *lymph glands* (nodes) (*see* Fig. 23) which produce antibodies, destroy bacteria and produce new white blood cells. These glands become swollen when the body becomes heavily in-

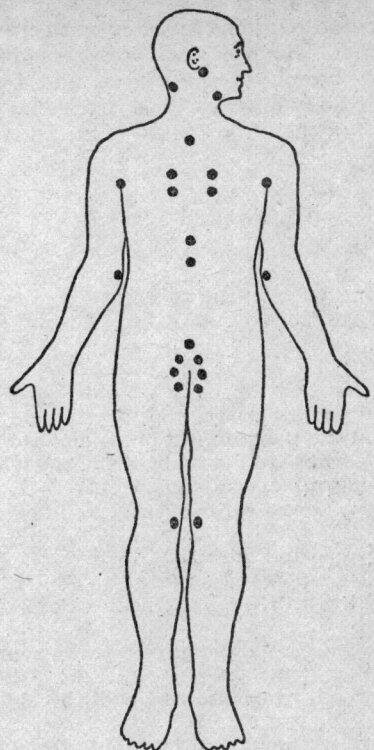

FIG. 23. Position of lymph nodes.

fected. Some diseases, *e.g.* mumps, are characterised by the swelling of these glands.

14. Function.

(*a*) To collect the liquid bathing the cells and return it to the blood system.

(*b*) To combat infection.

(*c*) To collect fatty acids and glycerol from the small intestine.

There is no special pumping mechanism to ensure the movement of the lymph. The necessary pressure is built-up by the accumulation of lymph in the tissue, and by the general movement of the body muscles. The semi-lunar valves ensure the one-way flow of the lymph towards the heart.

PROGRESS TEST 3

1. What are the two systems that make up the transport system in man? **(1)**

2. Name the four parts of the vascular system. **(2)**

3. Why is the vascular system called "closed" and "double"? **(2)**

4. Draw the vascular system of a man. **(2)**

5. Name the parts of the body supplied by (a) the hepatic artery, (b) the subclavian artery, and (c) the renal artery. **(2)**

6. Draw and label a diagram of the mammalian heart. **(3)**

7. In the heart, what are the functions of the following valves: (a) tricuspid, (b) mitral, (c) semi-lunar? **(4)**

8. In what way does cardiac muscle differ from striated muscle? **(4)**

9. What are systole and diastole? **(5)**

10. What is the average pulse rate of an adult? **(5)**

11. Give two ways to reduce the chances of suffering a heart attack. **(6)**

12. Give three structural differences between veins and arteries. **(7, 8)**

13. Excluding food, what else can pass through the capillary walls? **(9)**

14. How is oxygen transmitted in the blood? **(11)**

15. State four other substances transmitted in the blood. **(11)**

16. State three treatments for simple anaemia. **(12)**

17. State three functions of the lymphatic system. **(14)**

EXAMINATION QUESTIONS

1. Draw a diagram of the blood system of man. Trace the course of a red blood cell through it. Outline the changes that will take place in a red blood cell during its passage through the blood system.

2. Write an account of the functions of the blood.

3. With the aid of diagrams describe the structure of the heart. Outline the way in which blood is pumped through the heart.

4. Give accounts of: (*a*) anaemia; (*b*) heart attacks. Emphasise the methods of prevention and treatment.

5. Write short accounts of the following: (*a*) lymph; (*b*) serum; (*c*) leucocytes; (*d*) erythrocytes.

CHAPTER IV

THE SKELETON,
MUSCLES AND MOVEMENT

THE SKELETON

1. Introduction. The human skeleton is made of bone and cartilage.

(*a*) *Bone*. The microscopic structure of bone has been discussed already. At the centre of the bone is the soft marrow containing blood vessels, and this is the site of the

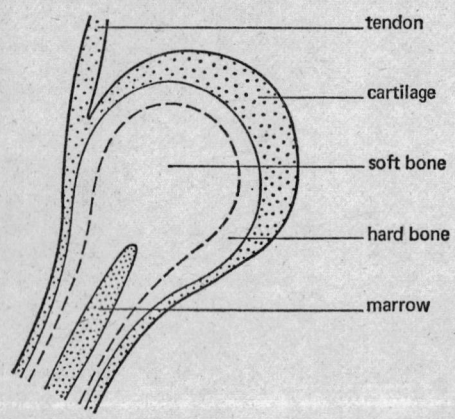

FIG. 24. End of large bone.

production of red blood cells. Outside this is a layer of soft, porous bone which is covered by a much denser and harder bone (*see* Fig. 24). Each bone is covered by a tough fibrous layer (*periosteum*).

The skeleton of a baby is cartilage in the early stages of its development. Calcium phosphate is laid down within the cartilage to make it hard (ossification), but just behind the ends of the bone, a zone of cartilage remains unossified, allowing the bone to grow. Ossification is not completed until the boy or girl has stopped growing.

(b) *Cartilage.* The microscopic structure of cartilage has been discussed already. Once ossification is complete, the only cartilage which remains in the skeleton is that found between and around the joints, in the nose, and in the lobes of the ear. Being elastic it functions as a shock-absorber, and as it is smooth, it allows the joints to move smoothly. Cartilage can grow, so that when it is worn away by the movement of the joints it can be replaced. Synovial joints are held together by a capsule of cartilage.

2. Functions of the skeleton.

(a) *Support.* During the process of evolution, animals have become larger, and moved away from water, which was their original habitat, onto the land. A small animal, *e.g. amoeba*, does not need a skeleton to support it, especially if it is living in water, which will support its weight. As animals became larger, a skeleton developed which assisted in locomotion and supported the internal organs. There are various forms of skeleton, *e.g.* a hard outer shell (*exoskeleton*) as found in crabs, lobsters, insects, *etc.*, and the internal bony skeleton (*endoskeleton*) of some fish, amphibians, reptiles, birds, and mammals. On land, the skeleton has to support the whole weight of the animal, so that the strength of the skeletal material, *i.e.* the bone, limits the size to which land animals can evolve. The largest animal, the whale, is aquatic.

(b) *Movement.* For an animal to move there must be muscles pulling on a rigid skeleton. The force exerted by these muscles is used to push against the external environment. Fish push against water, birds against the air, and man against the ground. When movement takes place, one part of the animal, *e.g.* the leg, moves in relation to the rest of the body, exerts a pressure on the external surface, and pulls the rest of the body up to itself. For this to take place, the rigid skeleton must be *jointed* and have *muscles attached to it* in such a way that they can bend the joints.

(c) *Protection.* The delicate organs of the body are protected by the hard skeleton. The skull protects the brain; the vertebrae protect the nerve cord; and the ribs protect the heart and lungs.

(d) *Red blood cells* are produced in the marrow of the larger bones and ribs.

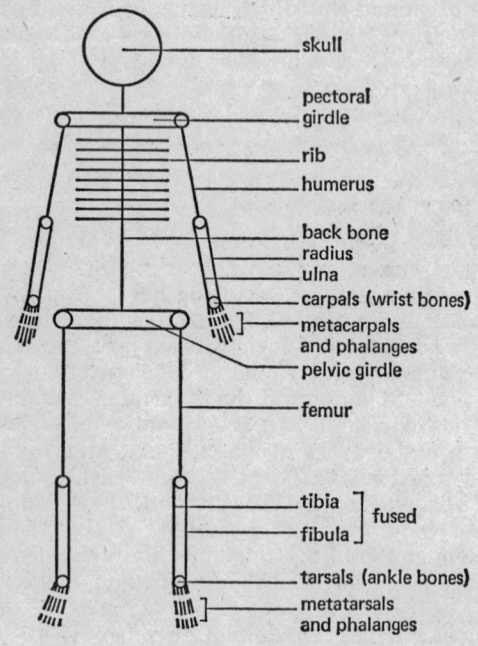

FIG. 25. Diagram of the skeleton.

3. Structure of the skeleton. The skeleton consists of two parts: (a) the *axial skeleton*, and (b) the *appendicular skeleton* (*see* Fig. 25).

(a) *The axial skeleton* includes the skull, the vertebral column, and the ribs.

The *skull* is made up of several paired bones joined together by zig-zag joints (sutures) which give the *cranium* great rigidity and strength. There are various holes in the cranium through which nerves and blood vessels pass. The most obvious ones are:

(*i*) The large foramen magnum at the base of the skull through which the spinal cord passes;

(*ii*) the openings to the middle ear on each side of the skull; and

(*iii*) the openings at the back of the eye orbit through which the optic nerves pass.

Beneath the cranium is suspended the lower jaw, which is jointed to the cranium, and with which it articulates during chewing and speaking.

There are twenty-six bones in the *vertebral column* (spine). They differ in structure as they have slightly different functions, but they have the same basic pattern. Each consists of a solid cylinder of bone (*centrum*) which articulate with each other forming a solid, flexible rod which supports the body. As one looks down the length of the body the size of the centra increases until they form the *sacrum* in which the centra are fused to each other and to the pelvic girdle. Thus the more posterior the bone, the greater the weight it is capable of supporting. Over the centrum of each bone is the *neural arch* forming the *neural canal*. The neural canals of the successive vertebrae form a tunnel which protects the spinal cord which runs through it. On the top of each neural arch is a *neural spine*, while on the side of the bone are flattened extensions called *transverse processes*. The neural spines and the transverse processes are for the attachment of muscles. On the neural arch and the transverse processes are smooth surfaces (*facets*) which allow the articulation of the bones. Between the centra of neighbouring vertebrae are pads of cartilage which act as a lubricant and shock-absorbers preventing jolting, particularly of the nerves and brain. Ligaments join successive vertebrae together so that the whole vertebral column forms a semi-flexible column (*see* Fig. 26).

(*i*) *Cervical vertebrae.* These are the vertebrae of the neck. There are seven of them. The first two are called the *atlas* and *axis*. They are not typical as they are modified to allow the movement of the skull. The atlas has no centrum. It is a heavy ring of bone with large facets anteriorly which articulate with the skull allowing it to move up and down. A membrane isolates the neural canal. The axis has a large *odontoid process* extending anteriorly from the centrum. This occupies the position of the centrum of the atlas, and being detached from the atlas acts as a pivot on which the atlas can turn from side to side. This arrangement of the two bones allows the movement of the skull in two planes.

The other five cervical vertebrae are small and light, giving flexibility to the neck and having little weight to support.

(*ii*) *Thoracic vertebrae.* There are twelve of these bones. They are relatively light. There is no great weight in the chest, so they have little weight to support. The neural spines

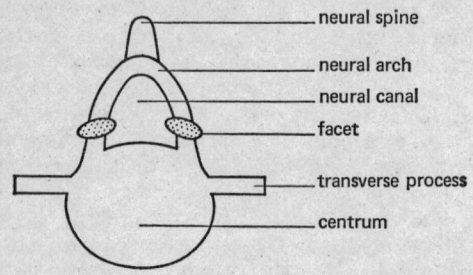

Fig. 26. Diagram of a vertebra.

are relatively long, and there are facets on the transverse processes and centra for the articulation of the ribs.

(*iii*) *Lumbar vertebrae.* There are five lumbar vertebrae. They are large bones with heavy centra and well-developed transverse processes on to which the back muscles are attached.

(*iv*) *Sacrum.* The sacrum consists of five fused sacral vertebrae through which there are lateral openings for the passage of nerves. Fusion of the bones gives strength, but flexibility is lost. The sacrum is fused to the pelvic girdle to form a large ring of bone which supports the weight of the intestine above it.

(*v*) *Caudal vertebrae.* These five bones form the *coccyx* which is a vestigial tail. Each bone is small, being a reduced centrum (*see* Fig. 27).

The *ribs* and *sternum* form the thoracic cage which protects the heart and lungs. There are twelve pairs of ribs
which articulate with the thoracic vertebrae. Ten of the
ribs are joined to the sternum by cartilage which completes
the cage. The lowest two ribs (floating ribs) are not attached

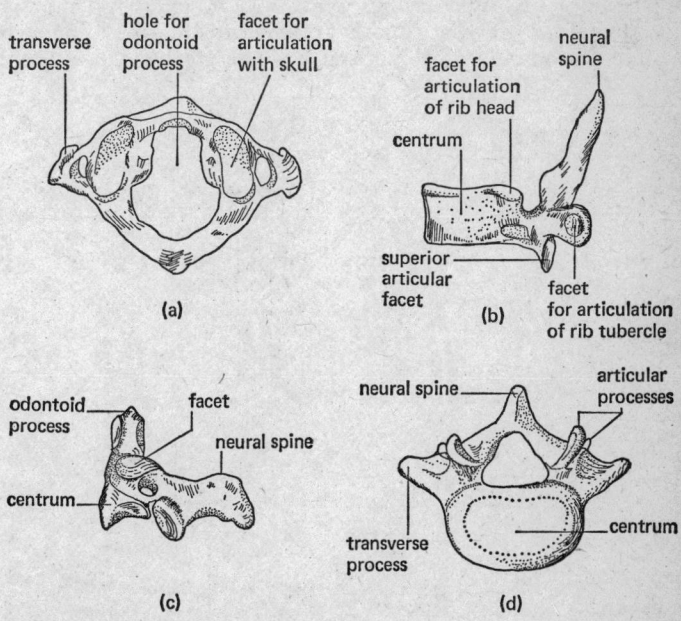

Fig. 27. Some bones of the vertebral column: (*a*) atlas vertebra
from above; (*b*) thoracic vertebra from side; (*c*) axis vertebra from
side; (*d*) lumbar vertebra from above.

to the sternum. Each rib consists of a ridged *shaft* with a
head and *tubercle* at one end. Intercostal muscles are
attached to the ridge, while the head and tubercle articulate
with the facets on the thoracic vertebrae. Contraction of
the intercostal muscles causes the movement of the ribs
during breathing, and this movement is made possible by

the articulation with the thoracic vertebrae and the presence of the flexible cartilage (*see* Fig. 28).

(*b*) *The appendicular skeleton* includes the limb girdles and the attached muscles.

The *pectoral girdle* or shoulder girdle is made up of the pair of *clavicles* or collar bones in the front and the flattened *scapulae* (singular *scapula*) behind. Each scapula articulates with a clavicle, and is attached to the ribs by muscles. On the outside of the scapula is the *glenoid cavity* for the

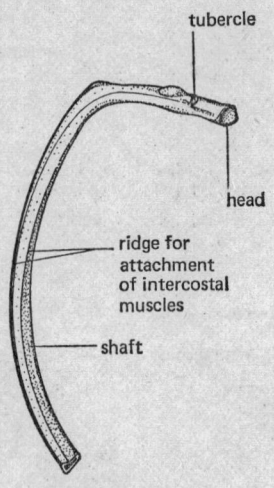

tubercle

head

ridge for attachment of intercostal muscles

shaft

FIG. 28. A rib.

articulation of the humerus of the arm, and the knob near the glenoid cavity is where the large biceps muscle of the arm is attached. This arrangement of bones allows for maximum flexibility, but this means that there is little strength in the girdle; however, since the arms do not have to support a great weight, this system is adequate.

The *pelvic girdle* (pelvis) is a massive structure consisting of two fused halves. Each half consists of three bones, the *ileum, ischium* and *pubis*, all of which are fused together. At the junction of these three bones is the deep hollow *aceta-*

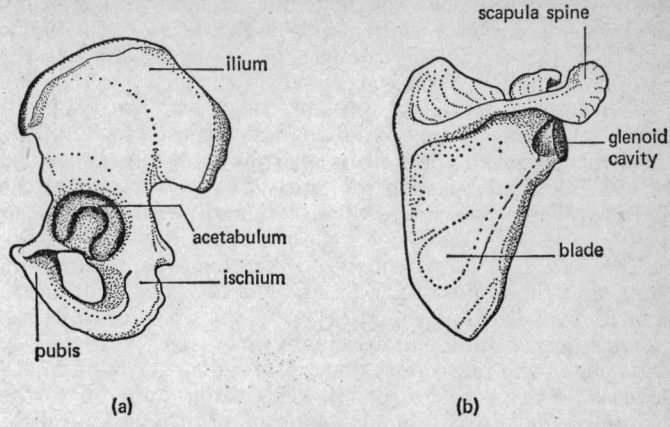

FIG. 29. Parts of the appendicular skeleton: (a) left half of pelvic girdle from side; (b) right scapula from above.

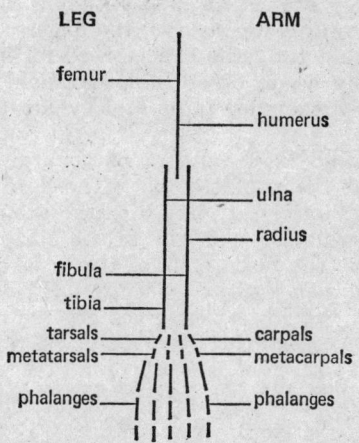

FIG. 30. Pentadactyl limb.

bulum in which the head of the femur articulates. The fusion of these bones means that there is no flexibility in the pelvic girdle, but there is the strength necessary to support the weight of the body conveyed to it by the legs (*see* Fig. 29).

The limbs of man, like those of other vertebrae, are built-up on the *pentadactyl* (five-fingered) pattern (*see* Fig. 30). The upper part of the limb is a single bone, the upper end of which articulates with the girdle. The lower part of the limb consists of two bones which articulate with the wrist or ankle bones and finally with the *phalanges* (toes or fingers).

The *fore limb* (arm) consists of the *humerus, radius* and *ulna,* and the bones of the hand. The *head* of the humerus articulates with the glenoid cavity of the scapula, while the *condyles* at the other end articulate with a socket formed by the ends of the radius and ulna. The radius and ulna in the lower arm are separate bones. The radius joins on to the "thumb side" of the hand, and can rotate over the ulna. This allows the hand to be rotated through 180°. Below the radius and ulna are the wrist bones (*carpals*) which articulate with the hand bones (*metacarpals*) which, in turn articulate with the finger bones (*phalanges*). The arrangement of the phalanges of the thumb is unique to man and monkeys. They are set at such an angle that the thumb can be bent towards (*opposed to*) the fingers. This arrangement means that the hand can be used for holding objects and not simply as an organ of locomotion. The opposed thumb is one of the most important evolutionary advances of man.

The *hind limb* (leg) consists of the *femur, tibia,* and *fibula,* and the bones of the foot. The head of the femur forms a strong universal joint with the acetabulum of the pelvic girdle, while the condyles at the other end form the knee joint with the fused ends of the tibia and fibula. At the knee joint is the knee cap (*patella*). Fusion of the tibia and fibula of the lower leg gives greater rigidity to the limb as a whole, but the foot cannot be rotated. This is unnecessary, as the foot is the pedestal on which the body stands. This is further emphasised by the fusion and relative sizes of the ankle bones (*tarsals*), foot bones (*metatarsals*), and toes (*phalanges*).

4. Types of joint. Between the bones there are two types of joint:

(*a*) *Immoveable joints*, *e.g.* between the bones of the skull.

(*b*) *Moveable joints*, which are classified by the type of movement they allow:

(*i*) *Vertebral joints.* Two bones are separated by a pad of cartilage, and joined together by ligaments. This allows a limited side-to-side movement.

(*ii*) *Pivot joints.* These allow a side-to-side movement,

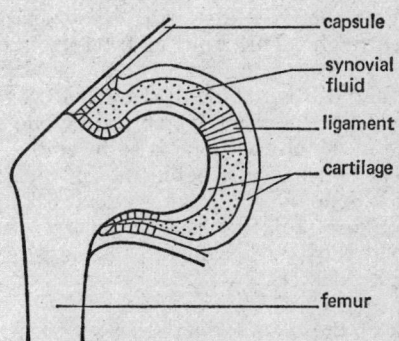

FIG. 31. Ball and socket joint.

e.g. between the condyles of the base of the skull and the atlas vertebra.

(*iii*) *Hinge joints.* These allow extensive movement, but only in one plane. The elbow is a good example of this. A side-to-side movement of the lower arm is impossible without involving the shoulder joint as well.

(*iv*) *Ball and socket joint.* This type of joint allows movement in more than one plane, *e.g.* the shoulder joint and hip joint. These are also examples of *synovial joints*. Instead of there being a pad of cartilage in these joints the lubrication is provided by a liquid—the *synovial fluid* which is secreted by the *synovial membrane*. The ends of the bone are covered by cartilage, and the bones forming the joint are held together by ligaments (*see* Fig. 31).

5. Growth of bone. In the early stages of development (before a baby is born) the skeleton is made of cartilage, and as development proceeds the cartilage becomes ossified. In the long bones two zones of cartilage are left near the ends so that the bone can continue to increase in length. When the individual is fully grown (20–25 years old) the remaining cartilage becomes ossified, and growth of the bone is no longer possible.

6. Bone fractures. Because it is rigid, bone is also relatively brittle and is liable to become broken by sharp knocks. The bones of young children are not fully ossified and are relatively flexible. If they become distorted, they snap, but do not break right through. This kind of fracture is called a *green-stick* fracture. A *simple fracture* is one in which the bone is broken, but the flesh is not pierced by the broken bone. In a *compound fracture* the flesh is pierced, while a *comminuted fracture* is one in which the bone is splintered.

Fortunately, bones are capable of self-repair, by forming a callus of new bone over a break. This is a long process, and for it to be successful the broken pieces have to be held together so that they cannot move. This is why a broken limb is put in a rigid plaster.

7. Diseases of the bone. *Tuberculosis* of the bone may be contracted by drinking infected milk. The tubercular bacillus attacks and destroys the bone.

Bone may become inflamed. This can be initiated by a knock, and then the inflamed tissue may become invaded by bacteria. The most serious form of this inflammation is called *osteomyelitis*, when the bone and marrow become inflamed.

MUSCLES AND MOVEMENT

8. Muscles. The muscles which make up the bulk of the body are *striated muscles*. The detail of their microscopic structure was described in I. Each muscle is made up of bundles of muscle fibres. Their contraction is under the control of the nervous system, the whole muscle being supplied with a large number of fine nerves.

The energy for contraction is derived from respiration, *i.e.*

from the release of energy from ATP. When a striated muscle contracts, the dark bands are seen to come closer together. Protein molecules between these bands lie along the long axis of the fibre, and during contraction they slide over each other, pulling the dark bands together. Energy from the ATP → ADP breakdown is used to bring about this "molecular sliding".

9. Attachment of muscles.

(a) Skeletal muscles are always attached to bones at both ends.

(b) The attachment is always across a joint, at least at one end, so that contraction of the muscle will cause movement of the joint.

(c) The attachment of the muscle to the bone is by a tendon.

10. Movement. Movement of a joint is brought about by the contraction of a muscle. As the muscle becomes shorter and thicker, it pulls on the relatively unextensible tendon, and consequently brings the two bones to which the muscle is attached together. Muscles always lie across joints in pairs which are opposed to each other (*antagonistic muscles*) so that as one muscle (*flexor*) contracts, the other (*extensor*) relaxes. Thus by the balance of tension between the two muscles across the joint, the bones (*e.g.* of a limb) can be held in any position.

11. Movement of the elbow. Movement across the elbow joint illustrates the above principle. The pair of opposed muscles are the *biceps* (down the front of the arm) and the *triceps* (down the back of the arm). The biceps is attached at one end to projections of the scapula, and at the other end across the elbow joint to the radius. Similarly, the triceps is attached at one end to the scapula, and at the other around the elbow to the *olecranon process* of the ulna. When the biceps muscle contracts (flexes) the triceps relaxes, and the fore-arm is raised. To straighten the arm the biceps relaxes, and the triceps contracts (*see* Fig. 32).

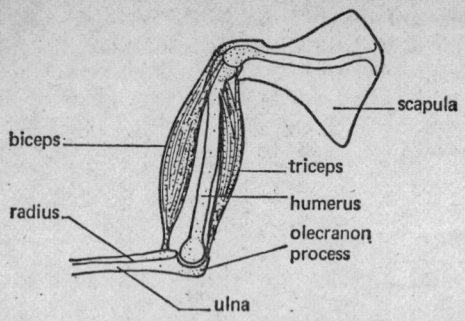

FIG. 32. Muscles of upper arm.

12. Posture. Man walks upright so that there is a constant force, the force of gravity, tending to displace the organs downwards. To overcome this force, and to hold the organs in their correct relative positions, the muscles are in constant use.

A correct posture is one in which the weight of the body is supported vertically over the ankles, along a line running perpendicularly from the ankle to the ear. When this posture is adopted, the body weight is supported directly down through the spinal bones and the leg bones.

If the shoulders are hunched, the head becomes thrust forward. In this position, balance can be maintained only by placing an undue strain on the back muscles, and by tilting the pelvis. In this position, it is impossible to achieve a full expansion of the chest muscles and disorders of the breathing system result.

Holding the head upright and thrusting the chest forward is frequently considered to be a smart way of standing. This posture throws the centre of gravity forward, so that there is a strain on the thigh muscles to maintain the posture. The abdominal muscles also become slack, leading to digestive and breathing disorders (*see* Fig. 33).

13. Tone and exercise. Even when the body is at rest, *e.g.* sitting or standing, the antagonistic muscles are in a slight tension to hold the body in position. This tension is muscle *tone*. To keep this tone, the muscles need exercise. Lack of

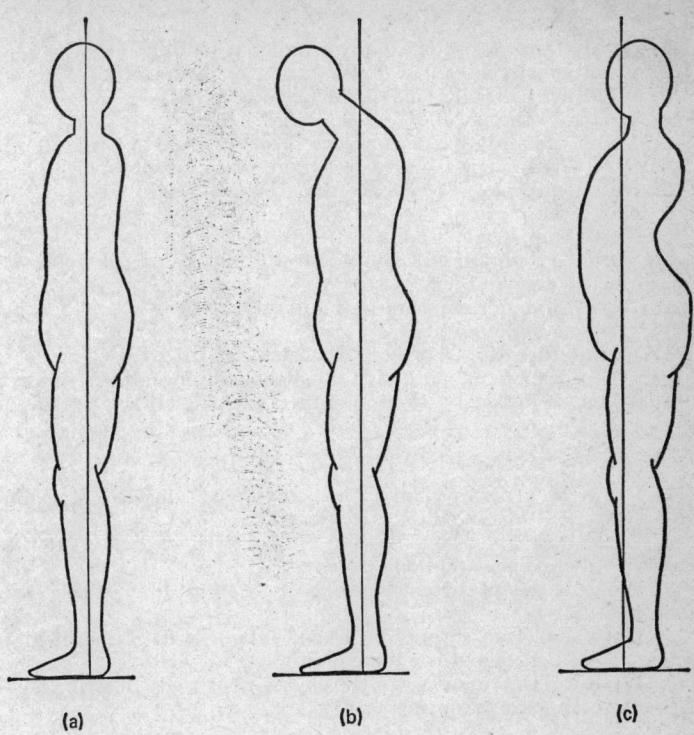

FIG. 33. Posture: (*a*) good posture; (*b*) round shoulders; (*c*) hollow back.

tone leads to flabby muscles, particularly the abdominal muscles, which in turn leads to a general physical and mental sluggishness.

Regular, relatively mild exercise is more beneficial than periodic violent exercise, especially to older people. A brisk walk every day will do a lot of good for the middle-aged office worker; a sudden violent game of squash may kill him through heart-failure.

PROGRESS TEST 4

1. Of what substance is a baby's skeleton made? **(1)**
2. What substance is laid down during ossification? **(1)**
3. What are the functions of the skeleton? **(2)**
4. What are the parts of the skeleton? **(3)**
5. Name the main parts of a vertebra. **(3)**
6. Give three particular features of an atlas vertebra. **(3)**
7. How many pairs of ribs are there in man? **(3)**
8. Name the parts of the pectoral girdle. **(3)**
9. Name the parts of the pelvic girdle. **(3)**
10. Draw a diagram of a pentadactyl limb. Label the bones as found in the arm and leg. **(3)**
11. Name four forms of moveable joints. **(4)**
12. Draw a synovial joint. **(4)**
13. Name three types of bone fracture. **(6)**
14. Draw the elbow joint with the attached muscles. **(9–11)**
15. What is meant by the tone of a muscle? **(13)**

EXAMINATION QUESTIONS

1. Draw the arm of a man with the attached muscles. Explain how movement of the elbow is achieved.
2. Write notes on the following: (*a*) growth of bone; (*b*) the function of cartilage; (*c*) bone diseases.
3. What is meant by good posture? Why is it necessary for good health?
4. Draw a section through a named hinge joint. Label it and outline the functions of the parts you have labelled.
5. What are the functions of the skeleton? Briefly describe how these functions are performed.

CHAPTER V

CO-ORDINATION

1. Introduction. An animal needs to detect changes in its environment and to co-ordinate the functions of the different organs of the body.

(*a*) *Detection of changes in the external environment* enables an animal to find food, protect itself, and find a suitable mate. Changes in light intensity and colour are perceived through the eyes; differences in sound through the ears; various chemicals (taste and smell) through small groups of sensory cells in the tongue and nose; and mechanical stimuli by receptors in the skin and mesenteries surrounding the internal organs. We are so used to *sense organs* functioning normally that we rarely realise how important they are. The absence of sense organs in the skin, for example, could lead to severe damage by burning through unwittingly picking up hot objects. Deafness could lead to being killed by a motor car that you did not hear coming.

(*b*) *Changes in the internal environment* are also perceived by sensory cells and elicit the appropriate response, *e.g.* hunger symptoms are caused by the collapse of the stomach wall which stimulates the nerve endings in it; changes in the carbon dioxide concentration in the blood are recorded by sensory cells in the brain.

(*c*) The realisation of the environmental changes would be of little value if there were no mechanisms whereby a *co-ordinated reaction* was elicited, which would be of benefit to the animal. For example, the sense organs in the skin recording that the object being touched was hot would be of no value unless the muscles reacted appropriately to move the hand away. Similarly, in a simple action, like picking up a pencil, there is a high degree of co-ordination. The position of the pencil is recorded by the eyes, and the information passed to the muscles of the arm and fingers, so that the precise movements needed can be performed—

there is no random groping, but a high degree of co-ordination between the movements of the different muscles involved.

(d) *A long-term co-ordination* is necessary during the growth and development of an individual. Each part of the body must develop in relation to all the other parts. A sustained reaction to a stimulus is required in certain cases, *e.g.* the continued production of enzymes while food is in the stomach.

(e) The fast stimulus ⟶ response type of action is under *nervous* control, while the slower, more sustained type of response to stimulus is controlled by *hormones*.

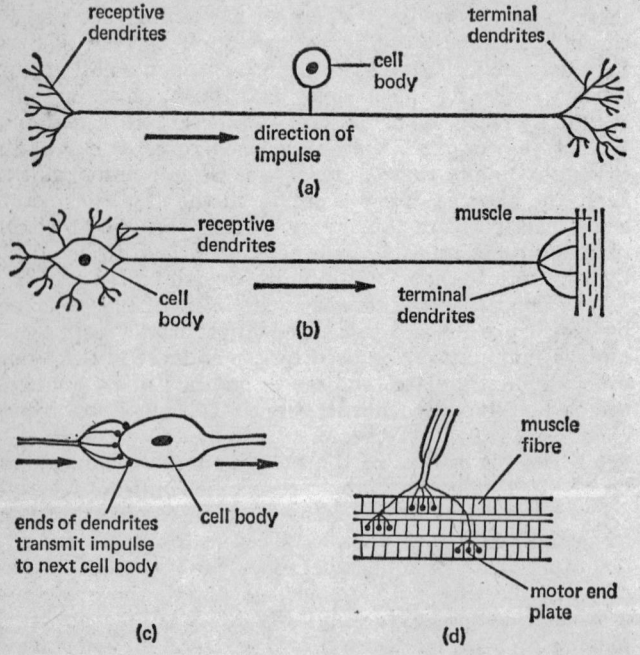

FIG. 34. Structures involved in the transmission of a nerve impulse: (a) sensory neurone; (b) motor neurone (*see* also Fig. 9); (c) synapse; (d) motor end plates.

THE NEURONE

2. Introduction. The nervous system is made up of *receptor cells* which record changes in the environment and *neurones* which carry nerve impulses from the receptor through various pathways to the *effector*. The effector is a muscle or nerve which will respond to the impulse (*see* Fig. 34).

3. Structure of the neurone. Each neurone consists of a *cell body* which contains a *nucleus*; the position of the cell body is different in motor and sensory neurones (*see* Fig. 34). From the cell body extends the long *axon* (it may be 1 m long in the spinal cord of a man), and in the motor neurone shorter branches called *dendrons* and *dendrites*. The axon is surrounded by a *myelin sheath* which is interrupted at intervals by the *nodes of Ranvier*. Around the fatty sheath is a fine membrane (*neurilemma*) in which can be seen the nuclei of the *Schwann cells* from which it is made.

Numbers of neurones are enclosed in a common sheath to form the nerves which run through the body. In a dissection they are seen as fine white silvery threads.

4. Functioning of the neurone.

(*a*) *Neurones will conduct impulses only in one direction.* Sensory neurones are sensitive to changes in the sensory cells, and conduct the impulses away from them through the axons. Motor neurones conduct impulses from the cell body along the axon to the effector which it stimulates. A nerve cell transmits the stimulus as an electrical impulse which travels along the axon. This impulse is similar to, but not the same as, an electric current. The electrical potential necessary for the conducting of the stimulus is generated by the passing of charged ions through the membrane of the axon wall. The myelin sheath acts as an insulator for each axon. The nodes of Ranvier are not insulated; this enables the impulse to jump from node to node rather than run the whole length of the axon. Such a system increases the speed of transmission along the axon.

(*b*) *Impulses pass from one neurone to the next through a synapse.* The adjacent nerve fibres are not touching, and the transmission across the gap is brought about by the chemical *acetylcholine*. Acetylcholine, produced at the end of the stimulated axon, affects the permeability of the membrane of the adjacent neurone and "triggers off" an impulse that

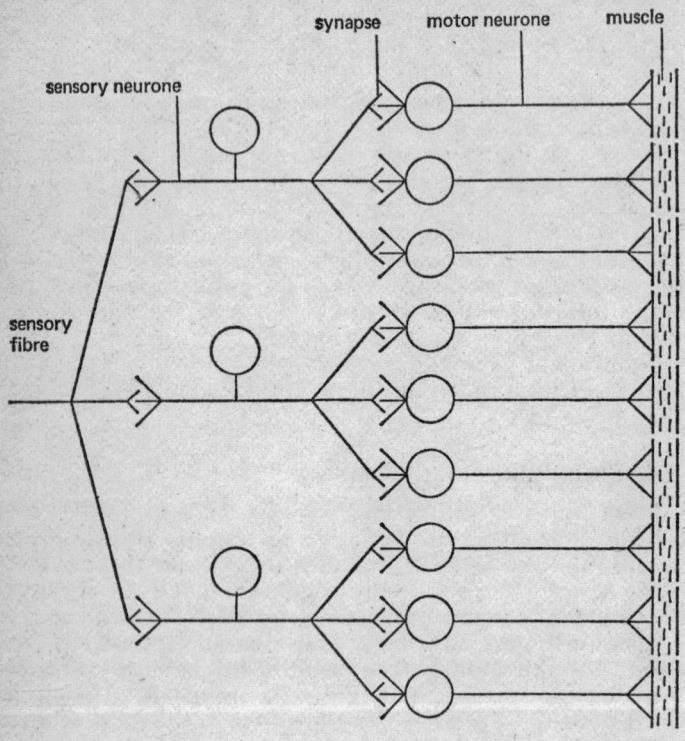

FIG. 35. Transmission of a single stimulus to several effectors.

can travel through it. As soon as the acetylcholine is produced, it is destroyed by the enzyme *choline esterase*. This prevents the continued stimulation of the nerve.

The presence of synapses means that several neurones can be connected with any other neurone, so that the stimulation of one receptor can excite one or many effectors. This idea is illustrated in Fig. 35.

(c) *The effector* is also stimulated by acetylcholine which is secreted by the *end plate* of the sensory nerves. End plates are similar in physiology to the synapses.

THE NERVOUS SYSTEM

5. Introduction. The nervous system of man consists of the *central nervous system* (CNS) and the *peripheral nervous system*. The CNS is the *brain* and *spinal cord*, while the peripheral nerves are a vast network of fine nerves which extend into every organ of the body. All the impulses pass through the CNS in which most of the cell bodies lie, so that it is within the CNS that the inter-connection between the nerves shown in Fig. 35 takes place.

6. The spinal cord. The spinal cord runs longitudinally down the length of the body, and is protected by the neural arches of the vertebrae. It is continuous with the brain. From it extend the spinal nerves which run through spaces between the vertebrae. Branches of the spinal nerves extend to all parts of the body.

Each spinal nerve enters the spinal cord as two branches (roots). The *dorsal root* contains sensory neurones and the *ventral root* motor neurones. These two roots fuse close to the spine, so that the spinal nerve contains both motor and sensory fibres, and is called a *mixed nerve*.

The spinal cord is covered by three membranes, the *dura mater* outside, the *arachnoid mater* beneath it, and the *pia mater* in contact with the nerve tissue. In transverse section the spinal cord has two distinct zones, with a central fluid-filled canal. The central zone is H-shaped. It contains the cell bodies and association neurones. This is the *grey matter*. Outside the grey matter is the *white matter* which consists largely of nerve fibres running longitudinally through it. The white colour is due to the myelin sheaths around the nerves seen in transverse section.

The spinal cord is the transmitting part of the nervous system. Nerve impulses transmitted from the sensory cells

are conveyed through the spinal nerves to the spinal cord and
are relayed from here, either directly to a receptor (as in the
case of a reflex), or to the brain.

7. Reflex action. A reflex action is one which does not
involve impulses being transmitted from the brain. They are
always rapid and automatic. In many cases the brain does not
even receive a stimulus after the reaction has taken place, so
that the person is not aware that the action has taken place.
For example, the production of saliva when food is placed in
the mouth, or the movement of the iris of the eye in varying
light intensities are reflex actions of which we are not aware.
Breathing is a reflex of which we are aware, but can do nothing
to control. (You can hold your breath for a certain time, but
ultimately one has to take in more air.) The lifting of the hand
to protect the head from a blow, putting out one's hands to
save oneself when falling, or the moving of the arms when
walking, are reflex actions of which one is aware and normally
does not control. They differ from the other two types in that
they can be controlled, *e.g.* you can walk with your arms held
to your side, and it is possible, although difficult, to walk
swinging the left arm with the left leg.

The classical demonstrable example of a reflex is the *knee
jerk*. If you cross one leg over the other, and tap the upper
leg below the knee cap, its lower part jerks out as a kick.

Most reflexes are *spinal reflexes*, *i.e.* the impulse travels from
receptor to effector through nerves in the spinal cord and the
brain is not involved, *e.g.* if the brain of a cat is destroyed,
movement of the right fore-leg initiates the movement of the
left hind leg.

Reflex actions are important in that they enable the normal
body functions to be performed without involving the brain.
This leaves the brain free to act as a memory bank, and to
initiate actions which involve some kind of thought.

8. Reflex arc. A reflex arc is the path taken by a nerve
impulse during a spinal reflex (*see* Fig. 36). A receptor is
stimulated and transmits an impulse along a sensory nerve
which is a part of the spinal nerve. The sensory branch enters
into the *dorsal root* of the spinal nerve and transverses the white
matter into the grey matter. In the grey matter the sensory

nerve forms synapses with *intermediate neurones* which in turn form synapses with motor neurones. These run out from the spinal cord through the ventral root, ultimately making connection with the effector (*see* Fig. 36).

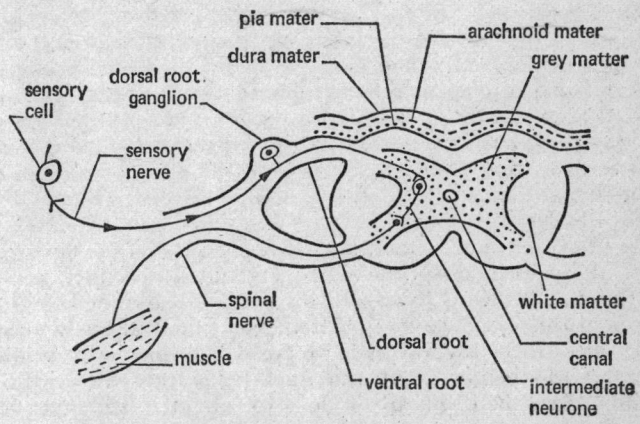

FIG. 36. Reflex arc.

9. Autonomic nervous system. In discussing reflex actions, reflexes which can be controlled were mentioned as were those which cannot be controlled. Those which can be controlled involve *striped* muscle, while those which cannot be controlled, or are unconscious, involve smooth muscle (*e.g.* the iris of the eye) or glands. The smooth muscles and glands are under the control of a subsidiary nervous system called the *autonomic nervous system.*

The autonomic nervous system consists of two nerve cords running parallel to the backbone, with ganglia at intervals. Each ganglion is connected to a spinal nerve, so that impulses passing through the autonomic system can pass into a reflex arc.

Functionally there are two systems within the autonomic system. They are the *sympathetic nervous system* and the *parasympathetic nervous system.* These two systems work in opposition to each other, the sympathetic system preparing the body in stress conditions, *e.g.* ready to fight or run away (*see below* the hormone adrenalin in **40**) while the parasympathetic system causes the various organs to relax, *e.g.* when asleep, or when food is being digested.

10. Conditioned reflex. In the normal reflex action the stimulus and response are related, *e.g.* the secretion of enzymes by the stomach wall when food passes into the stomach. But it is possible to elicit a reflex response to an apparently unrelated stimulus. This is called a *conditioned reflex*. Many of our day-to-day activities are conditioned reflexes. The characteristic of a conditioned reflex is that the normal channel of the reflex stimulus is altered by the process of learning.

The classic experiments of the Russian biologist, Pavlov, illustrate this point. A normal reflex path is one in which the sight of food stimulates the salivary glands to produce saliva, *i.e.* the mouth begins to water. Pavlov presented food to dogs in which this response was elicited, but simultaneously a bell was rung. After several days of presenting food in this way he rang the bell without the food being present, and the ringing of the bell caused the dogs to salivate. The dogs had *learnt* to associate the ringing of the bell with the presence of food, and the salivating response had been conditioned to result from the originally unrelated stimulus of the ringing bell.

11. Voluntary actions. These actions involve the brain, more particularly the *cerebral cortex* which is the centre of intellectual activity. A voluntary action is one which is "thought about". It is the result of a positive mental process, and a decision is made. This type of action is remembered, and is based on learning from previous experience. When communication between individuals is highly developed, as it is in man, the previous experience can be that of the individual, or that of another individual the result of which is communicated by language.

Let us consider the example of turning on a tap. A person wants some water. He has learnt, either by being told, or by trial and error, that if the tap is turned in an anti-clockwise direction, water flows out of the pipe. Having this information stored in the memory, the brain is now capable of deciding to send the necessary impulses to the appropriate muscles in the arms and hands so that the tap is turned. Notice how this reaction differs from a reflex. Simply because the tap is there the person does not turn it on. If he did the action would be a reflex. Tap-turning is only performed as a conscious act as a result of previous experience, to acquire a desired result.

THE BRAIN

12. Function of the brain. The behaviour of the lower animals is based on reflexes and instinct. These animals are capable of only limited amounts of learning, if any. For example, it is impossible to train a worm to carry out a complex behaviour pattern. Man has developed highly efficient sense organs at the anterior end of the body (head). This stage in evolution necessitates a corresponding enlargement of the CNS to form a *brain*. Once the evolution of the brain had begun, it continued, getting relatively bigger to become:

(a) *A correlation centre* for impulses received by the various sense organs.

(b) *A store of information* received by the receptors (memory).

(c) *A centre from which impulses are transmitted*, some on the basis of information instantly received from the sense organs, and some on the basis of memory.

(d) *The centre of reason and intelligence.*

13. Development of the brain. The brain develops as an enlargement of the spinal cord. During the early stages, the enlargement is into three distinct but joined chambers; the *fore-brain, mid-brain* and *hind-brain*. (*See* Fig. 37.) Later the fore-brain becomes greatly enlarged to form the paired *cerebral hemispheres*. Their relative enlargement is so great, that they become folded backwards over the mid-brain and hind-brain, so as to partially enclose them (*see* Fig. 37).

14. The fore-brain. The fore-brain receives impulses from the nasal organs. These impulses are transmitted by *cranial nerves* which connect the brain directly with the organs concerned.

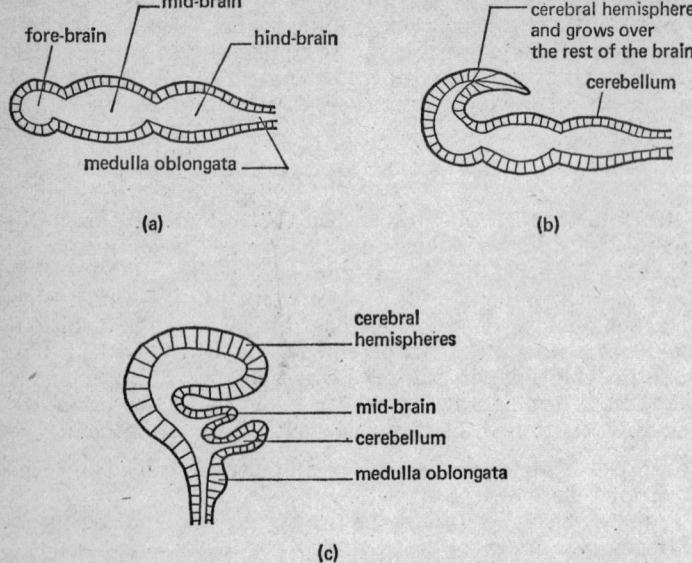

FIG. 37. Development of the brain: (*a*) fore-brain, mid-brain and hind-brain; (*b*) formation of cerebral hemispheres; (*c*) folding of cerebral hemispheres.

The *cerebral hemispheres* are developed from the fore-brain. Located in the cerebral hemispheres are *motor areas* which initiate movement of various parts of the body (*see* Figs. 38, 39).

The cerebral hemispheres are also the centres of memory and learning, so it is not surprising to find centres (*association centres*) in the surface layers (*cerebral cortex*) which co-ordinate the responses to the variety of stimuli received. It is within the association centres that the "decision" is made on the reaction to a particular set of circumstances in the light of the variety of impulses received, and what has been remembered from previous experience.

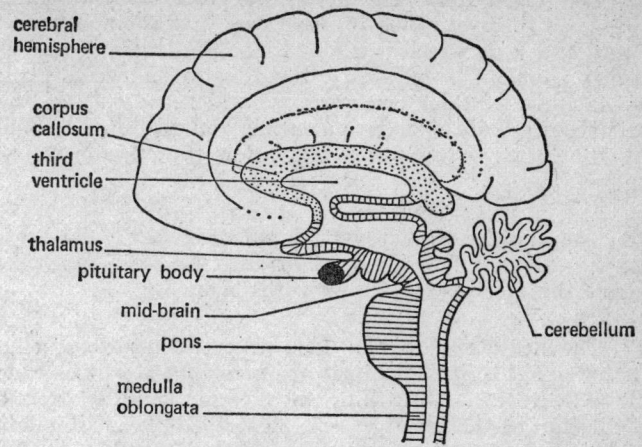

FIG. 38. Section through brain.

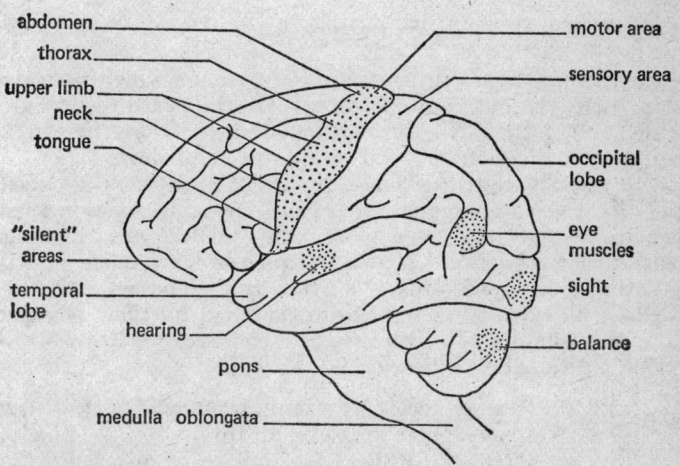

FIG. 39. Areas of the brain.

15. The mid-brain. The mid-brain receives the cranial nerves from the eyes. Closely associated with the mid-brain is the *pituitary body* which is divided into the *thalamus* and the *pituitary gland*. The pituitary gland is an endocrine gland. The thalamus contains nerve centres which control the "survival activities" of the body, *e.g.* eating and drinking, sleeping, body temperature and osmo-regulation (functioning of the kidneys).

16. The hind-brain. The hind-brain receives cranial nerves from the ears and skin. It develops into the *cerebellum* which controls the muscles concerned with balance.

17. The medulla oblongata. This lies between the cerebellum and the spinal cord. Within it are *involuntary* centres which control heart beat, breathing movements, and contraction and dilation of the blood vessels. The medulla is stimulated by a high concentration of carbon dioxide in the blood, which causes it to transmit impulses which accelerate the heart beat and movement of the intercostal muscles.

RECEPTORS

18. Introduction. The receptors are organs which perceive changes in the external environment. The sense organs are connected to the CNS through synapses, so that there is a complete connection between receptor and effector.

The sensory cells are single, in small groups, or grouped together as sense organs. Generally speaking sense organs respond to only one type of stimulus. There are apparent contradictions to this because the precise mechanism of stimulating many sense organs is not fully understood.

The sensory cells or organs are classified by their position or by the stimulus to which they are sensitive.

The positional classification is:

    (*a*) *Exteroceptors*, receiving stimuli from outside the body.
    (*b*) *Enteroceptors*, receiving stimuli from within the body.
    (*c*) *Proprioceptors*, which respond to changes in the tension of the muscles.

The functional classification depends on the response to the stimulus, *i.e.*:

   (*a*) *light*;
   (*b*) *sound*;
   (*c*) *gravity* (*balance*);
   (*d*) *taste*;
   (*e*) *smell*;
   (*f*) *pressure* (*touch*);
   (*g*) *temperature*.

Pain is caused by the over stimulation of any of these organs or cells. This has a great biological value as it prevents damage to cells and organs. For example, without the "safety valve" of pain, one would hold a red-hot poker with impunity, causing severe damage to the tissues of the hand. Similarly, without pain, an excessively bright light would cause irreparable damage to the retina of the eye.

Intensity of sensation is recognised by:

   (*a*) *The number of sensory cells stimulated*, *e.g.* if a small area of the skin is touched by a hot object only a few sensory cells are stimulated, while if a large area is touched more sensory cells are stimulated. Similarly, a light touch on the surface of the skin with a needle would stimulate only the surface cells, but a greater pressure would stimulate the cells below as well.

   (*b*) *Variation in the sensitivity of sensory cells.* Some sensory cells do not respond until the stimulus is intense (*see* Fig. 40).

19. Pressure receptors. These are situated in the skin and in the mesenteries of the gut. They are unevenly distributed, *e.g.* the tips of the fingers and lips have a greater density of these receptors than elsewhere.

The touch receptors are found singly. One form of touch receptor is the *pacinian corpuscle*. This is made up of a series of separate layers, so that in section it looks like an onion that has been cut through. Pressure on the corpuscle distorts these separate layers and sets up an impulse in the nerve which supplies it.

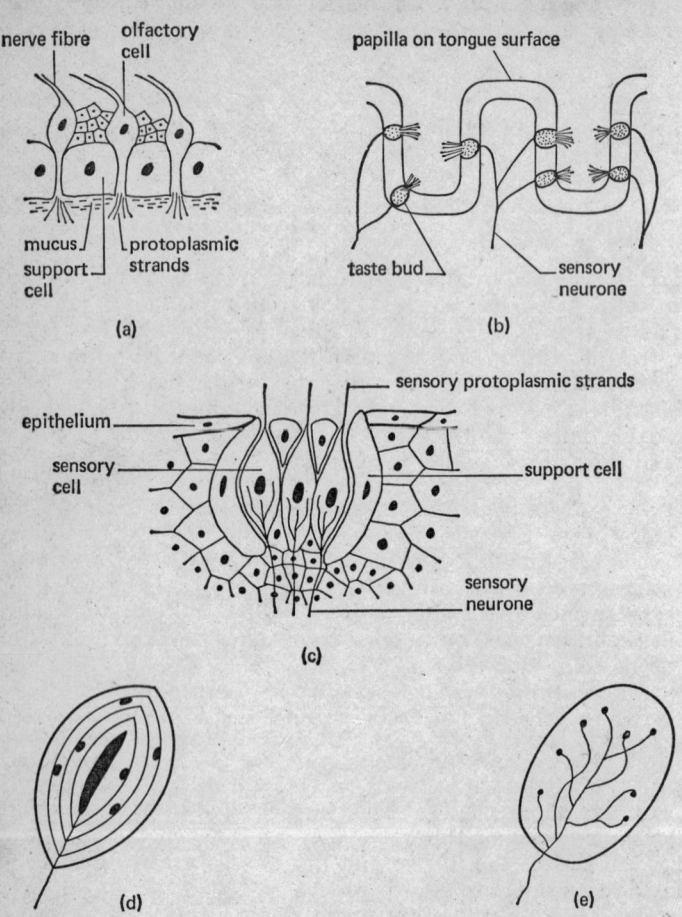

FIG. 40. Some sensory cells: (a) olfactory organ; (b) position of taste buds; (c) taste bud; (d) pacinian corpuscle. It lies beneath the dermis and detects pressure; (e) Meissner's corpuscle. It lies under the epidermis, close to hairs and it detects light pressure.

20. Temperature receptors. These are distributed through the skin, with greater concentrations in the lips, back of the hands, and elbow. Hence the method of testing the temperature of a baby's bath with the elbow. There are two types of temperature receptors, the *oval corpuscles*, which detect increases in temperature, and the *bulbous corpuscles*, which detect decreases in temperature.

21. Smell receptors. These are confined to the olfactory mucous membranes in the nasal cavity. The sensory cell lies within the mucous membrane and from it extend fine protoplasmic threads into the layer of mucus covering the membrane. Nerve fibres from the sensory cells are connected through synapses with the olfactory lobe of the brain. These cells respond to different chemicals *in solution* (chemoreceptors) stimulating the protoplasmic hairs. Minute particles of the substance being smelt are breathed in through the nose or pass back through the naso-pharynx from the mouth, and have to dissolve in the mucus before they stimulate the hairs. Many of the substances that we taste are, in fact, smells. Both the senses of smell and "taste" are diminished when we have a cold. This is because the mucous membranes become covered with a thick layer of mucus and the dissolved chemicals are not able to contact the protoplasmic hairs.

22. Taste receptors. These are chemoreceptors situated on the tongue. The upper surface of the tongue is roughened by the presence of numerous projections (*papillae*) along the sides of which are situated small groups of chemoreceptors called *taste buds*. From each cell of the taste bud a small protoplasmic hair extends into the surrounding liquid. The taste buds are sensitive to only four sensations, *sweet*, *sour*, *salt*, and *bitter*. Any other "tastes" are smells. The taste buds sensitive to "sweet" and "salt" are at the tip of the tongue, those sensitive to "bitter" at the back, and those sensitive to "sour" at the sides.

THE EAR

23. Introduction. The ear is the organ of *hearing* and *balance*. In fishes it is solely an organ of balance, and has evolved into an organ of hearing in terrestrial vertebrates (*see* Fig. 41).

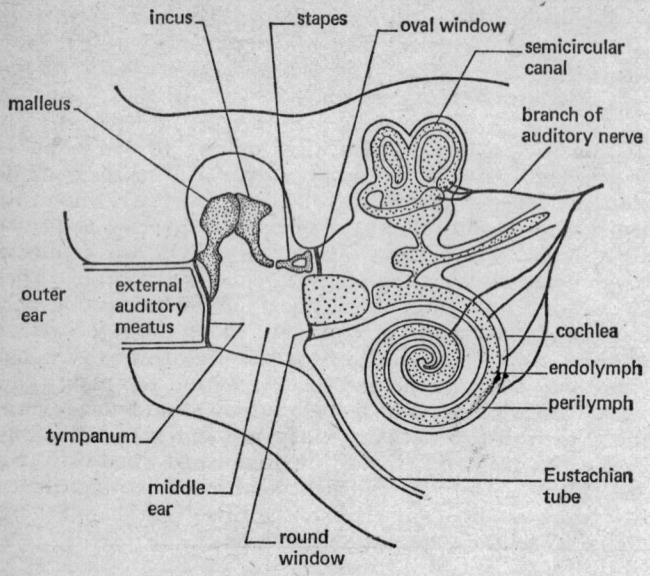

FIG. 41. Section through ear.

24. Structure. The ear has three distinct regions, the *outer ear*, *middle ear*, and *inner ear*.

(*a*) *The outer ear.* The external part of the outer ear is the cartilagenous *pinna*. This functions as a trumpet to collect the sound waves and direct them into the sensory part of the ear. It is not very important in man, but in other animals, *e.g.* a dog or horse, the pinna is large and can be turned in the direction of the sound. The necessary muscles are present in man as vestigial structures, and some people can use them to move the pinna to a limited extent. From the pinna a tube (*external auditory meatus*) leads inwards, and terminates in the ear drum (*tympanum*). The walls of the external auditory meatus contain wax glands. The wax they produce helps to protect the ear and lubricate the tympanum. Hairs which line the external auditory meatus to some extent exclude dust and small insects.

(b) *The middle ear.* The middle ear is a small cavity in the bone. It contains air, and opens externally into the pharynx by the *Eustachian tube.* Across the middle ear there are three small bones suspended by ligaments. These bones, the *malleus, incus* and *stapes,* are shaped and suspended so as to act as a series of levers which increase the force of the vibrations transmitted across them from the tympanum to the *oval window.* The presence of the Eustachian tube ensures that the pressures on each side of the tympanum are equal, *i.e.* atmospheric pressure. This allows the tympanum to oscillate freely when it is agitated by sound waves. The Eustachian tube is normally closed, but opens during swallowing or yawning. Blocking of the Eustachian tube, *e.g.* when we have a cold, may lead to temporary deafness, due to the inability of the tympanum to move freely. A similar effect occurs when the external pressure changes rapidly, *e.g.* when ascending in a lift, or taking-off in an aeroplane. Under these conditions, yawning usually restores hearing. During yawning or swallowing, the ears sometimes "pop". This because the Eustachian tube opens, equalising the internal and external pressures.

(c) *The inner ear.* The inner ear is filled with fluid. It is closed by two membranes, the *oval window* and the *round window.* This arrangement means that any vibration transferred to the oval window can pass through the fluid and be taken up by the round window without the vibration being reflected back through the fluid. The inner ear is a complex membranous structure called the *membranous labyrinth.* It is filled with fluid (*endolymph*). The membranous labyrinth is suspended in liquid (the *perilymph*) which is contained in a cavity of the bone called the *bony labyrinth.*

There are two main parts to the inner ear, which although connected are functionally distinct. They are the *semicircular canals* which are concerned with balance, and the *cochlea* which is concerned with hearing.

25. Balance. The semicircular canals are concerned with registering the position of the head and so maintain balance.

At the base of the semicircular canals are the *sacculus* and *utriculus*. The sensory structures are arranged mutually at right angles, each terminating in a swollen *ampulla* which contains a receptor. The ampulla receptor is a gelatinous cone extending into the cavity of the ampulla and attached at the end to sensory cells which are connected to nerve fibres. Within the utriculus are patches of sensory hairs on which are embedded small particles of calcium carbonate (limestone). When the head moves the inertia of the fluid in the semicircular canals tends to keep it still, so that the ampulla sensory organs are distorted and strain the sensory hairs. This impulse is transferred to the nerve. The calcium carbonate particles are denser than the fluid in the ear, so always tend to press vertically downwards. If the head is tipped to one side, the particles, instead of pressing downwards, will pull at an angle. This distortion is also transmitted to the nerves as a stimulus. The relative positions of the semicircular canals ensures that the pressure exerted by the fluid on the sense organs is different for each position of the head.

26. Hearing. Sound waves pass through the outer ear and strike the ear drum which begins to vibrate. These vibrations are amplified during transmission through the ear ossicles to the oval window. Vibration of the oval window sets up similar vibrations in the fluid of the cochlea which cause the *tectorial membrane* of the *organ of Corti* to vibrate against the hairs of the sensory cells. These vibrations are translated into impulses which are transmitted to the brain. The tectorial membrane in different parts of the cochlea is stimulated by notes of different pitch. The part nearest the round window is stimulated by high-pitched notes, and that nearer the end by notes of low pitch. The lower of the three canals of the cochlea (*scala tympanum*) is in contact with the round window so that the vibrations set up in the fluid can be dissipated (*see* Fig. 42).

The human ear is capable of detecting notes of frequency from 20 to 20,000 cycles per sec. Other animals, *e.g.* dogs and bats, can hear notes of higher frequencies. Human beings can distinguish the tones of different notes, *e.g.* you can distinguish the difference between the same note played on a piano and a violin. The mechanism of this is not clear.

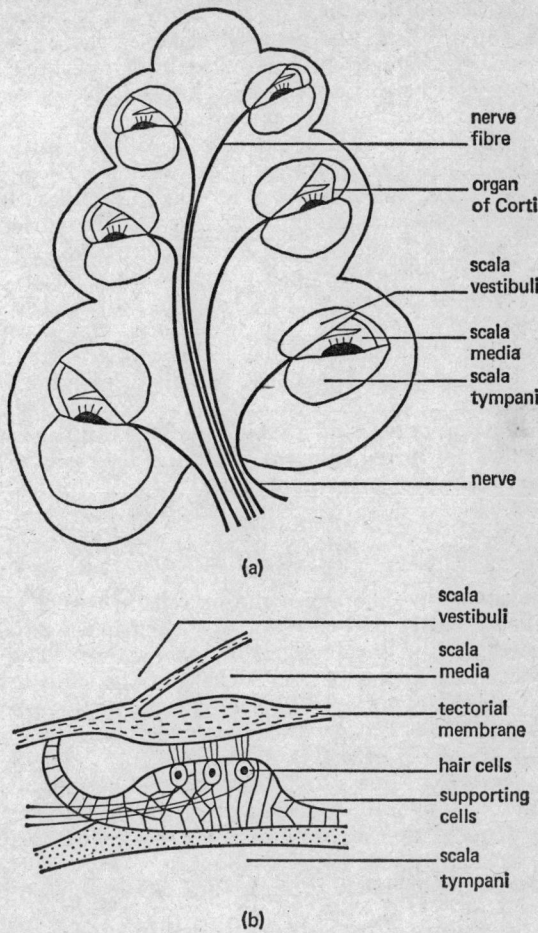

(a)

(b)

FIG. 42. Hearing: (a) section through cochlea; (b) organ of Corti.

27. Diseases of the ear.

(a) *Mechanical damage.* The ear drum can be damaged by pushing objects into the ear, *e.g.* cleaning the wax with a match stick, or small children accidentally pushing beads, modelling clay, etc. into the ear. Only a doctor or nurse should insert anything into the ears.

(b) *Inflammation of the middle ear.* This frequently causes ear ache and discharges from the ear. It may be caused by direct infection, accumulation of wax, or indirectly as a consequence of complaints such as measles and scarlet fever.

(c) *Infection of the mastoid bone.* This bone lies immediately behind the ear. It has a cavity which opens into the middle ear. An infection of this porous bone causes severe ear ache and discharge. The whole area can become inflamed. This infection can be fatal, but treatment with antibiotics has made this a less serious disease than it was previously.

(d) *Blockage of the Eustachian tube.* This can be caused by mucus formed during a heavy cold, and may cause inflammation and ear ache.

THE EYE

28. Introduction. The eye contains cells sensitive to light. These cells are situated at the back of the eye in the *retina*, and are connected to the optic nerve through synapses. The paired eyes are situated in the sunken sockets of the skull to which each is attached by six muscles. Movement of the eye in the socket is brought about by the antagonistic contraction and relaxation of these muscles.

Each eye sees an image individually, and the impulses are co-ordinated in the brain to form a single image. The eyes of a man are situated in the front of the head so that each eye perceives a slightly different image of the same object. This gives a three-dimensional (*stereoscopic*) picture of the object. The stretching of the eye muscles, which contain stretch receptors, is also translated into impulses which are transmitted to the brain. The stereoscopic vision and the extent of the stretching of the eye muscles make it possible to estimate distance. That both eyes are necessary to estimate distance can be demonstrated by trying to thread a needle with one eye shut.

Although stereoscopic vision has many obvious advantages, it has the disadvantage of there being no all-round vision. Acute vision is only possible in a very small area directly in front of the eyes, while there is *peripheral vision* through an angle of just over 100° from the direct line of sight. You will appreciate this if you hold a pencil at arm's length directly in front of you and, keeping your arm extended, swing it round. Keep looking straight ahead, and notice the position of your arm when the pencil disappears from sight. Other animals, particularly herbivores, have the eyes situated at the side of the head. They do not have the advantage of stereoscopic vision, but can see all around them.

29. Structure.

(a) *Conjunctiva.* This is a thin transparent layer which is continuous with the lining of the eye lid, and lies over the surface of the eye.

(b) *Sclerotic.* A tough, semi-elastic tissue enclosing the eye (the white of the eye). It covers the whole of the surface of the eye, but is transparent in front. Here it is called the *cornea.* The pressure of the fluids inside against the sclerotic maintain the shape of the eye ball.

(c) *Cornea.* As the cornea is curved and transparent, it functions as a lens and helps to focus the incoming rays of light on the retina.

(d) *Choroid.* This is a fine layer of tissue densely suffused with blood vessels which supply the eye ball with food and oxygen. It is densely pigmented, looking nearly black. This reduces the amount of light reflected within the eye ball. If the light was reflected the eye would receive several images of the same object. The dense pigmentation of the choroid and retina can be seen as the black pupil of the eye.

(e) *Retina.* The light-sensitive cells are confined to the retina. There are two types of light-sensitive cells called *rods* and *cones* which are differently shaped. The cones are sensitive to light of different wave length (different colours) and the rods are sensitive to light of low intensity. The cones contain a pigment *iodopsin* and the rods contain *rhodopsin*.

Both these dark purple pigments (formerly called "visual purple") are bleached by light. The bleaching of the pigment sensitises the nerves attached to the cells and an impulse is transmitted to the optic nerve and hence to the brain. Immediately after bleaching the pigments are regenerated.

(f) *Fovea* (*Yellow spot*). This lies directly opposite the centre of the lens. It is a specialised area of the retina containing only cones. This area of the retina gives the most detailed interpretation of the image. Light entering directly through the centre of the lens is focussed on the fovea and is seen clearly, while light entering from elsewhere is focussed on the rest of the retina giving peripheral vision.

(g) *Blind spot*. This is the area of the retina occupied by the end of the optic nerve. There are no sensory cells here so no impulses are generated when light falls on it.

(h) *Vitreous humour*. This is a thick jelly-like fluid filling the back of the eye. It is clear and refracts the light, thus being part of the system which focusses the light on the retina. The pressure of the vitreous humour and aqueous humour against the semi-rigid sclerotic keep the eye ball in shape.

(i) *Aqueous humour*. This fluid occupies the front cavity of the eye. It is less viscous than the vitreous humour, but has the same functions. In addition it is the source of food and oxygen for the conjunctiva, cornea and lens. None of these structures contains blood vessels.

(j) *Lens*. The crystalline lens is clear and biconvex. It is semi-rigid and surrounded by a fibrous capsule. The curvature of the lens can be altered by tensions set up in the ciliary body, thus altering its focal length.

(k) *Suspensory ligament*. The suspensory ligaments hold the lens in position, attaching it to the ciliary body. It is virtually inelastic.

(l) *Ciliary body*. This is an extension of the choroid which contains circular muscles, *i.e.* the muscle fibres run around the lens. You can visualise this as the lens lying in the centre of an elastic band (the ciliary body) and attached to it by non-elastic pieces of string (suspensory ligaments).

(m) *Iris*. The iris is a further extension of the choroid, extending in front of the lens. It is pigmented, giving the eye its colour, *e.g.* blue or brown eyes, and contains circular muscles.

(*n*) *Tear glands*. Although not strictly part of the optical system, the tear glands play an important part in seeing. They lie over the top of the eye ball and continually secrete a dilute solution containing sodium chloride, sodium hydrogen carbonate and an enzyme which destroys bacteria. This

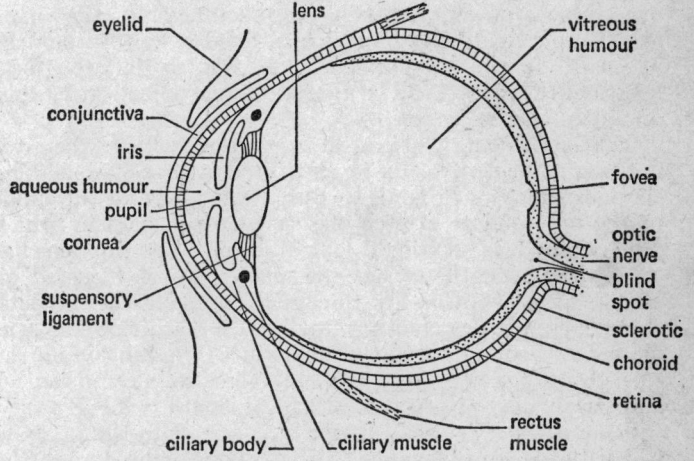

FIG. 43. Horizontal section through right eye.

fluid (tears) keeps the surface of the eyes moist and washes away dust particles, etc. The tears are collected from the surface of the eye into the tear duct, in the inner corner of the eye. This leads into the nasal cavity.

(*o*) *Eyelids*. The eyelids protect the eye. Blinking is a reflex which is induced by sudden movement towards the eye, strong light, or foreign matter coming in contact with the eye ball. Blinking also moves tears over the surface of the eye (*see* Fig. 43).

30. Function.

(*a*) *Seeing*. Light from an object enters the eye and is focussed by passing through the cornea, lens, aqueous humour and vitreous humour. An inverted, real image is formed on the retina. This image is upside down. The retinal cells send impulses to the brain through the optic nerve. In the optic centre of the brain, the inverted image is reversed, so that we "see" the image the right way up. Also in the brain, an impression of distance and size of the object is formulated.

(b) *Accommodation.* Accommodation is the ability of the eye to focus on objects at varying distances. If the whole system was rigid, only objects at a fixed distance would be in focus. In the eye the lens is flexible so that the focal length of the system can be altered. Thus, objects at various distances can be focussed.

The lens is contained in an elastic capsule which, when it is not under tension, tends to reach its smallest size, making the lens short and rounded. But the pressure of the fluids of the eye tend to stretch the suspensory ligament which pulls on the lens, making it long and thin. The thin lens has a long focal length so that the relaxed eye is focussed on objects at a distance. For the eye to focus on close objects, the ciliary muscles contract, and as they are circular their diameter becomes smaller. This reduces the tension on the suspensory ligaments and allows the lens to become shorter and fatter, *i.e.* to have a smaller focal length (*see* Fig. 44).

Frequently the lens cannot accommodate sufficiently, which leads to defective vision. Long-sightedness (the inability to focus on near objects) is called *hypermetropia* or, in the case of old people, *presbyopia*. In the latter case, the lens becomes tough and loses its elasticity. In either case, the light entering the eye is focussed so that it would form a clear image behind the eye ball. Long-sightedness is corrected by wearing converging (convex) lenses. Short-sightedness (*myopia*) is the inability to see distant objects clearly. It is frequently caused by doing a lot of close work, *e.g.* reading or embroidery. In this case, the lens forms a clear image of a distant object within the eye ball, and it is corrected by wearing spectacles with diverging (concave) lenses. Long-sightedness and short-sightedness may be due to the abnormal shape of the eye ball. If the eye ball is too small light from the lens is focussed behind the retina and long-sightedness results; similarly, if the eye ball is too large,

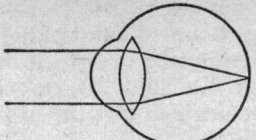

Light from object brought into focus on yellow spot

(a)

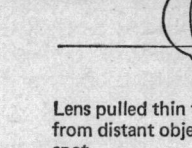

Lens pulled thin to focus light from distant object on yellow spot

(b)

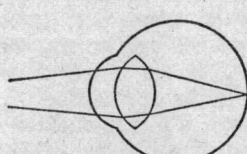

Lens allowed to become thicker to focus on near object

(c)

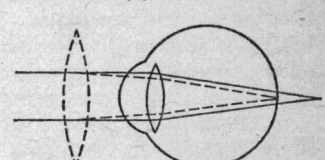

Rays which would normally be focussed outside the eye are converged by a convex lens

(d)

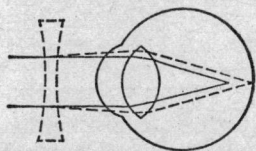

Rays which would normally be focussed inside the eye are diverged by a concave lens

(e)

Fig. 44. The focussing of the eye and correcting abnormal vision: (a) focussing of eye; (b) focussing on distant object; (c) focussing on near object; (d) correction of long sight; (e) correction of short sight.

light is focussed within the eye ball causing short-sightedness.

(c) *Control of the amount of light entering the eye.* This is brought about by the iris. In bright light, the circular muscles in the iris contract reducing the aperture of the pupil, while in dim light the radial muscles contract widening the pupil. In either case, the optimum amount of light is allowed to reach the retina, avoiding damage to it by bright light, and reduced sensitivity in dim light. The movement of the muscles of the iris is a reflex action initiated by the amount of light falling on the retina.

31. Conditions causing malfunctions of the eye.

(a) *Cataracts*. The lenses in the eyes of old people may become hard and opaque. This is called a cataract. It can be treated surgically by removing the lens. The person then needs spectacles to see. As the eye can no longer focus, close- and long-distance, or bifocal glasses are needed.

(b) *Corneal abnormalities*. Irregularities in the surface of the cornea affect the focussing of light on the retina. These are called *astigmatisms*. They can be corrected by lenses. Severe irregularities in, damage to, or diseases of the cornea can make it opaque. When this happens it is possible to remove it surgically, and replace it with the cornea from a dead person.

DRUGS

32. Introduction. Most drugs have an effect on the nervous system; some increase its activity (*stimulants*), while others (*depressants*) reduce its activity.

Many drugs, *e.g.* alcohol and opium, have been used by human beings since time immemorable to produce a euphoric state, and the dangers of using them were not realised. Other drugs, *e.g.* "pep pills", were first used medicinally, and as such are of great value, but when used indiscriminately they are dangerous.

The chief danger of drugs is that they can be habit-forming. A person may well try a drug for fun, and become addicted to it remarkably quickly. This addiction begins simply as a desire to repeat the pleasant sensation, but it leads to a condition in which the body metabolism is altered so that it *needs* the drug to function properly. This is the true state of addiction.

Drugs taken in excess invariably are harmful, both physically and mentally.

A person under the influence of drugs can be a menace to himself and to other people. He may behave irresponsibly and even criminally, *e.g.* being drunk in charge of a car.

Drug addiction is very difficult to cure, so the best policy is to avoid drugs at all costs, except when prescribed by a doctor.

33. Alcohol. Alcoholic drinks have been taken throughout history. Often they were (and still are in some countries) the only safe drink to take when water supplies were unfit for human consumption. We do not know how many cases of alcoholism there were in history, but now it is potentially a serious social problem, *e.g.* there are some $4\frac{1}{2}$ million alcoholics in the U.S.A.

The immediate effect of taking even a small amount of alcohol is an impairment of the nervous system. Reactions become slower, and there is a loss of judgment. Hence, the critical tests (breathalysers) used on motorists involved in road accidents if they are suspected of drinking. The cerebellum is one of the areas of the brain affected so that muscular co-ordination is impaired.

Alcohol is one of the few food substances that are absorbed directly through the stomach wall. In a matter of a few minutes after drinking, it is present in the blood vessels of the stomach.

Other effects of taking alcohol are:

(*a*) *A dilation of the blood vessels of the skin.* This is why alcohol should never be given to a person in a state of shock or suffering from cold. The dilated blood vessels cause excessive heat loss from the surface of the skin.

(*b*) The flow of blood to the skin *deprives the other organs of their normal blood supply* which may result in a malfunction of the kidneys.

(*c*) Prolonged heavy drinking can cause *permanent damage to the liver* (*cirrhosis*).

(*d*) Prolonged heavy drinking can affect the brain to such an extent that the person has *hallucinations* (*delirium tremens*).

An alcoholic cannot control his craving for drinking large amounts of alcohol. This contrasts with the heavy drinker,

who can. Alcoholism is a disease. The person suffering from
it is usually unable to "face up to life", due to a feeling of
insecurity, and seeks escape in drinking. The disease is basically psychological and needs proper medical care. This usually
involves treatment with drugs. There is a world-wide voluntary organisation called Alcoholics Anonymous which does a
great deal of good work, based on group therapy, in curing
alcoholics.

34. Depressants. These substances decrease basic body
metabolism including heart beat and respiration. Depressants
include:

(*a*) *Aspirin* and related compounds. These are valuable
for the relief of pain and reducing temperature. If taken in
excess they can cause delirium, sickness and an irregularity in heart beat. There is some evidence that they may
cause stomach ulcers. Aspirin is cheap and readily available, but should not be taken for minor headaches, etc.

(*b*) *Opium* is derived from the juice of the opium poppy.
In its raw state its use is confined nearly exclusively to
Oriental countries. It is smoked, causing little damage.

(*c*) *Morphia and heroin* are derived from opium, but are
much stronger. Medicinally they are used as pain-killers,
but are very easy to become addicted to. They cause a
complete breakdown in personality. The addict becomes
unable to sleep, depressed, and shuns the company of other
people. He is unable to hold down a job, and so is unable
to obtain the money to pay for the drugs which he must
have. This leads to economic breakdown which ruins him
and his family. These drugs are administered in solution by
a hypodermic injection.

(*d*) *Tranquillisers* are used to reduce stress and worry.
They are valuable when taken under medical supervision,
but can be dangerous if taken in excess.

(*e*) *Barbiturates* are one of the main constituents of sleeping tablets. An over-dose may cause defects of the nervous
system such as loss of memory and paralysis, and can be
fatal.

35. Stimulants. These increase the basic body metabolism,
affecting heart beat and respiration. Excess of the stronger

ones causes delusions, hallucinations, and may induce the person taking them to perform violent and criminal acts.

(a) *Caffeine* is found in small amounts in tea, coffee, and cocoa, but one would have to drink large amounts of these beverages for the drug to have any serious adverse effects. Medicinally, caffeine is used as a stimulant, frequently to counteract the depressant after-effects of pain-relieving drugs like aspirin.

(b) *Cannabis, Indian hemp, marijuana, hashish* are derived from the Indian hemp plant. The drug may be ingested or smoked in cigarettes (reefers). Large doses cause confusion and delirium. The addict may behave depravedly and commit criminal acts.

(c) *Amphetamine, Drinamyl, benzedrine* are the bases of "pep pills". These are used medicinally to counteract depression, but are often taken (having been procured illegally) to give one a "lift". Students sometimes use them when studying for examinations. One can easily become addicted to these drugs, leading to irresponsible behaviour. As the sale of "pep pills" is illegal except on a doctor's prescription, the addict is an easy victim for the drug "pusher".

DISEASES OF THE NERVOUS SYSTEM

36. Poliomyelitis. This disease attacks the brain and spinal nerves, causing paralysis of the muscles supplied by them. If the respiratory muscles are affected, death may result.

The disease is caused by a virus which is present in the throat. It is spread from one person to another by coughing and nasal discharges, and by food and drinks. Flies can be carriers of this disease.

People of all ages, except the very old, are susceptible to the disease, but it tends to affect the young more readily.

It has an incubation period of 5–21 days, so that contacts should be quarantined for 21 days, and an infected person isolated for 42 days. The disease is endemic, and in the past occasionally reached epidemic proportions.

The symptoms are a slight fever and a general drowsiness. Paralysis (it if occurs) usually sets in early, but muscular symptoms may be confined to an aching back and limbs.

The treatment involves complete rest, but is a matter for expert attention. An "artificial lung" may be used if the chest muscles are affected.

Poliomyelitis can be prevented by treatment with a vaccine made from weakened (*attenuated*) strains of the virus. This is called the *Sabin vaccine*. It is given orally to babies 6–12 months old—three doses at one-monthly intervals. The disease is notifiable.

37. Meningitis (meningococcal fever, cerebrospinal fever).

This is an inflammation of the membranes surrounding the brain and spinal cord. It is caused by a coccus bacterium. The symptoms are a fever and general restlessness. The disease is infectious.

Treatment is by sulphonamide drugs and penicillin. Bromides may be used to relieve the pain. Pressure on the nerve is relieved by removing the spinal fluid through a lumbar puncture.

38. Encephalitis.

This is caused by a virus infecting the brain. The infection is spread from an infected eye or ear, or from brain damage.

The symptoms are headache, sickness, possibly blindness and coma. The disease is infectious. Permanent brain damage may result, so that the patient may have to be confined to an institution.

THE ENDOCRINE SYSTEM

39. Introduction.

The endocrine system consists of a number of *ductless glands* which produce *hormones*.

The hormones are secreted directly into the bloodstream and usually affect organs or tissues at some distance from the site of secretion. Frequently, the secretion of one hormone will initiate or prevent the secretion of another, so that there is a mutual integration of the functioning of the glands.

The function of the hormones is to co-ordinate the activities and development of the body. In this their function is similar to that of the nervous system, but it differs in that the hormones function more slowly, and their effects are longer lasting. Adrenaline is one of the fastest-acting hormones, but it functions more slowly than a nerve impulse. Other hormones, *e.g.* those affecting growth, are produced over a number of years; their effect is gradual and prolonged.

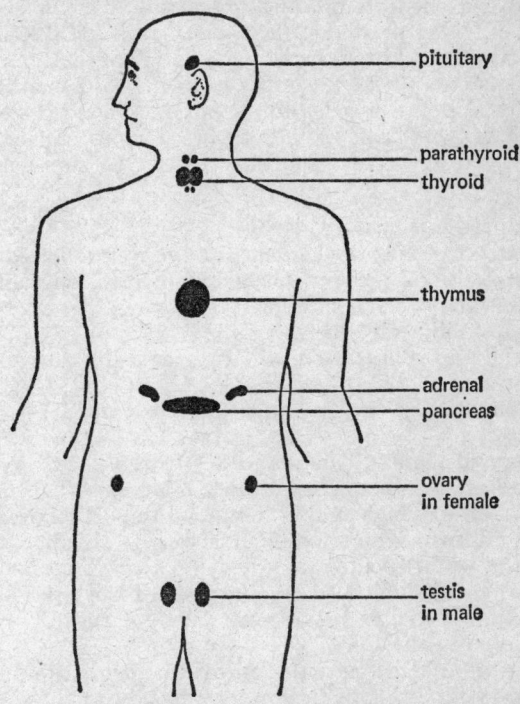

FIG. 45. Positions of the glands.

40. Adrenal (suprarenal) glands. The pair of adrenal glands lie immediately on top of the kidneys (*see* Fig. 45). They are divided into two distinct zones, the outer *cortex* and the inner *medulla*. The adrenal cortex produces a number of hormones:

(*a*) One which regulates the *activity of the kidney*, thus controlling the concentration of sodium and potassium ions in the blood:

(b) one which controls the *formation of sugars from proteins*;

(c) *sex hormones*.

The adrenal medulla produces *adrenalin*. This is frequently called the "fight or flight" hormone. It is produced under stress conditions and reinforces the function of the sympathetic nervous system. Its main effect is to improve the tone and activity of skeletal muscles to enable more effective activity. Several organs are affected to ensure a more efficient blood supply to the muscles. The effects may be summarised as follows:

(a) Increased heart beat;

(b) blood vessels to the heart and muscles dilate;

(c) muscles of the gut and skin contract, releasing more blood for the skeletal muscles;

(d) breathing rate increases;

(e) the pupils of the eyes dilate, increasing the sensitivity of the eyes;

(f) the tone of the skeletal muscle improves.

41. Thyroid gland. This gland lies in the neck, just below the larynx. It produces the hormone *thyroxine*. This hormone contains iodine, which makes iodine an important element in the diet. In areas where iodine is lacking in the diet (particularly in the water) people may develop a swelling in the neck (*goitre*) which is due to an enlargement of the thyroid gland. The addition of iodine to table salt prevents this. The disease is rarely seen in the U.K.

Thyroxine affects the rate of metabolism, especially the oxidation of foods. This indirectly affects growth. A normal production of the hormone is essential for growing children to ensure their normal physical and mental development. Lack of thyroxine in children results in mental retardation and disproportionate development of the head and hands. This condition is called *cretinism*. It can be avoided by adminis- tration of thyroxine extracted from animal thyroid over a long period (possibly years). In adults, a lack of thyroxine also causes a slowing of metabolism, but then it shows as a general mental and physical laziness, and increase in weight and a thickening of the skin (*myxoedema*). Excess thyroxine gives the person an excitable temperament; he is frequently anxious, nervous and irritable. The person tends to become thin, with slightly protruding eyes. The thyroid may become enlarged to produce an exophthalmic goitre.

42. Parathyroids. These are four small glands lying behind the thyroid. They produce the hormone *parathormone* which controls calcium and phosphorus metabolism.

43. Islets of Langerhans. These are diffuse masses of tissue in the pancreas which produce the hormone insulin.

Insulin controls the oxidation of glucose by the tissues and the storage of glycogen in the liver and muscles. It is being secreted into the blood constantly so that the amount of sugar in the blood is maintained at the level necessary for the activities of the tissues. If the production of insulin is reduced, the tissues cannot utilise sugar as a source of energy, and it is not stored. This leads to an excess of glucose in the blood, and this is removed by the kidneys, being revealed as excess sugar in the urine. Sugars not being available, the tissues utilise proteins and fats as sources of energy. This leads to a wasting of the muscles and other tissues, and the accumulation of toxic bi-products, especially aceto-acetic acid. Metabolism is reduced, resulting in coma and death. This condition is called *diabetes*. Diabetes can be treated by injection of insulin, which was originally extracted from the pancreases of animals, but can now be synthesised.

44. Testes. The interstitial cells of the testes (*i.e.* the cells between the tissues producing spermatozoa) become active during adolescence and produce the hormones *testosterone* and *androsterone*. These stimulate the development of the seminal vesicles and prostate gland, and the secondary sexual characters, *e.g.* deepening of the voice, beard and pubic hairs.

45. Ovaries.

(*a*) The cells of the *Graafian follicles* produce *oestrogen* which:

(*i*) Produces the secondary sexual characters during adolescence, and maintains them;

(*ii*) brings about a thickening of the uterine wall in preparation for the implantation of a fertilised egg;

(*iii*) prevents the pituitary gland (*see* 47) from producing the follicle stimulating hormone, ensuring that only one egg matures during each oestrus cycle.

(*b*) The *corpus luteum* produces *progesterone* which:

(*i*) Causes enlargement of the uterus in preparation for pregnancy;

(*ii*) inhibits the production of the pituitary sex hormones;

(*iii*) stimulates the development of the placenta;

(*iv*) stimulates the growth and development of the mammary glands, and initiates milk production.

(A full description of the female oestrus cycle is given in VIII.)

46. Mucous membranes of the intestine.

(*a*) Those of the stomach produce *gastrin* which stimulates the glands of the stomach wall to produce the gastric juices.

(*b*) Those of the duodenum produce *enterogastrone* (secretin) which inhibits the production of the gastric juices by the stomach.

The production of both these hormones is initiated by the presence of food in the particular organ, so that the gastrin production is stimulated by the presence of food in the stomach, but is inhibited as the stomach empties and the food passes into the duodenum.

47. Pituitary gland.

This gland is situated immediately under the brain. It is really two glands:

(*a*) The *anterior pituitary*, which developed from the roof of the mouth;

(*b*) the *posterior pituitary*, which has developed from the base of the brain.

It produces several hormones, listed below, but its basic function is to control the production of hormones by other glands. The hormones it produces are:

(a) *Pituitrin*—controls growth, especially of the long bones. An excess results in giantism, and a deficiency in dwarfism.

(b) *follicle stimulating hormone* (*FSH*)—stimulates the Graafian follicles to produce oestrogen, and a similar hormone stimulates the interstitial cells of the testes to produce testosterone.

(c) *prolactin*—stimulates the mammary glands to produce milk.

(d) *vasopressin*—influences the osmoregulatory processes of the kidney.

(e) *adrenotropic hormone*—stimulates the adrenal cortex.

(f) *thyrotropic hormone*—stimulates the thyroid gland to produce thyroxine.

(g) *glucagon*—increases the conversion of glycogen to sugars. This has the opposite effect to insulin.

(h) *oxytocin*—stimulates the muscles of the womb.

RELATION BETWEEN THE NERVOUS SYSTEM AND ENDOCRINE SYSTEM

48. Co-ordination. Both the nervous system and the endocrine system are concerned with the co-ordination of the activities of the body. Many endocrine glands are activated by nervous stimuli, *e.g.* the production of adrenaline is stimulated by an impulse from the sympathetic nervous system, which has in turn been stimulated by the brain.

The production of one hormone can stimulate another gland to produce its hormone.

49. Removal of stimulus. The removal of the stimulus which "triggers off" the whole complex of responses has a similar effect in preventing the final response, so that the system will return to normal.

50. Abnormalities. Remember that in the great majority of people the nervous system and endocrine glands function in perfect harmony. The abnormalities mentioned above are infrequent, and should not be over-emphasised.

PROGRESS TEST 5

1. Name the two systems responsible for co-ordination.   (1)
2. Name the types of cells which (a) record changes in the environment, (b) transmit nervous impulses, and (c) respond to the impulse.   (2)
3. Draw and label a nerve cell.   (3)
4. What is the function of the synapses?   (4)
5. Name the parts of the nervous system.   (5)
6. What do you understand by a "mixed nerve"?   (6)
7. Define a reflex action.   (7)
8. What is a spinal reflex?   (7)
9. What organs are controlled by the autonomic nervous system?   (9)
10. Define a conditioned reflex.   (10)
11. Define a "voluntary action".   (11)
12. State four functions of the brain.   (12)
13. What is the function of the cerebral cortex?   (14)
14. What is the function of the association centres?   (14)
15. What are the functions of the thalamus?   (15)
16. What is the function of the cerebellum?   (16)
17. What are the functions of the medulla oblongata?   (17)
18. What do you understand by (a) exteroceptors, (b) enteroceptors, and (c) proprioceptors?   (18)
19. How is the intensity of sensation recognised?   (18)
20. Where are pressure receptors found?   (19)
21. Where are the greater concentrations of temperature receptors?   (20)
22. Where are the smell receptors located?   (21)
23. Name the four types of taste receptor.   (22)
24. What are the functions of the ear?   (23)
25. What are the regions of the ear?   (24)
26. Name the bones in the middle ear.   (24)
27. What is the function of the Eustachian tube?   (24)
28. What are the two main parts of the inner ear?   (24)
29. Make a diagram of the eye and label the parts.   (29)
30. How is the amount of light entering the eye controlled?   (30)
31. Give two harmful effects of taking drugs.   [32]
32. Give the two types of drugs which affect the nervous system?   (32)
33. State four adverse effects of drinking alcohol.   (33)
34. State adverse effects of taking (a) morphia, (b) cannabis, (c) benzedrine.   (34, 35)
35. Name three diseases of the nervous system.   (36, 37, 38)

36. Where are the following hormones produced? State their functions: (a) thyroxine, (b) insulin, (c) adrenalin. **(40, 41, 43)**

EXAMINATION QUESTIONS

1. If a hot plate is picked up and dropped, trace the path of the nerve impulses from the instant of picking-up the plate to the time it is dropped.

2. Draw a diagram to show the structure of the human eye. Explain how an image is focussed on the retina. How can long sight be corrected?

3. Write notes on the following: (a) the Eustachian tube; (b) semicircular canals; (c) the cochlea, taking particular care to describe how they function.

4. Write an essay on the subject "Drugs as a hazard to health".

5. Compare and contrast the nervous system and hormones as co-ordinating systems. Briefly outline the functions of the following hormones: (a) thyroxine; (b) oestrogen; (c) insulin.

CHAPTER VI

HOMEOSTATIC MECHANISMS

1. Introduction. One of the main problems of survival for a human being is the maintenance of the internal conditions of the body in a constant state in an ever-changing external environment. At a more fundamental level, this means that the immediate environment of the individual cells must be kept constant to enable them to function properly. The cells can withstand variations in the environment for a short time, but if it is changed radically for any length of time, they are killed. This means that:

(*a*) The cells must be bathed in a solution of a constant chemical composition;

(*b*) The body must be kept at a constant temperature.

The systems that are concerned with keeping the body at a constant level are called *homeostatic mechanisms*. As you study each mechanism you will see that they are stimulated to function by a change in the internal environment, which they then correct, and when they have restored the optimum condition they cease to function. But, as the internal environment is constantly changing, the homeostatic mechanisms are being constantly brought into operation. This situation is like driving a car (the body) at a constant speed where the engine is being accelerated or braked depending on the slope of the road (the environment). The homeostatic mechanisms are the brake and the accelerator.

BREATHING

2. Introduction. The breathing mechanism is necessary to ensure efficient *gaseous exchange* between the body and the atmosphere. It is responsible for the introduction of oxygen into the body and the removal of carbon dioxide. The mechanism is homeostatic in that the rate of breathing is governed by the needs of the tissues for oxygen. (This will be discussed later.)

Gaseous exchange takes place through the lungs, which are enclosed within the surface of the body. Enclosure of the exchange surfaces is necessary to avoid their being dried-out. Gases can only pass through cell membranes in solution, there-

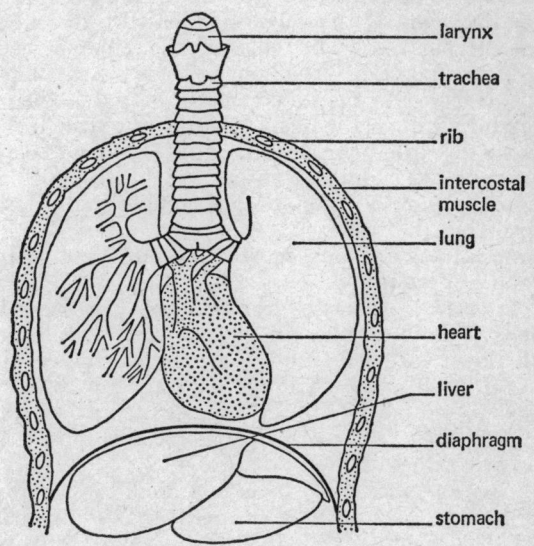

FIG. 46. Section through thorax.

fore the surfaces involved have to be moist. In an active animal, like man, the surfaces also have to be large to allow sufficient oxygen and carbon dioxide to diffuse through them in a given time. Exposure of a wet surface of such a large area to the atmosphere would involve such a rapid loss of water that the man would die of desiccation. Because the lungs are enclosed inside the body, a pumping mechanism is necessary to draw in and expel the air (*see* Fig. 46).

3. Structure. The breathing system consists of a pair of lungs which are provided with air passages which open to the exterior through the mouth and nose.

(a) *The nose* opens to the exterior by a pair of *nostrils* which lead into the nasal passages.

(b) *Nasal passages.* The surface of the nasal passages is increased by the fine *scroll bones* which are covered with an epithelium, some of the cells of which secrete mucus while others are ciliated. The cilia of the ciliated cells beat towards the nostrils. The mucus traps dust particles and organisms from the air, this mucus is wafted to the exterior. Air passing through the nasal passages is warmed. Thus the air that is breathed through the nose is warm and partially decontaminated before it enters the lungs. Air breathed through the mouth is not decontaminated and not so efficiently warmed.

The nasal passages are separated from the mouth by the *hard* and *soft palates*.

(c) *Trachea.* Air passes from the nasal passages into the trachea. This is a tube which runs from the back of the mouth down into the thorax, where it divides into two, each branch supplying a lung. Throughout its length, the trachea is supported by incomplete rings of cartilage. These hold the tube open when the pressure inside it alters during breathing.

(d) *Larynx* (*voice box*). This is a modification of the top of the trachea. Its opening (the *glottis*) is protected by the flap of the *epiglottis*. When food is swallowed it is closed, but when there is no swallowing movement, it is open. Inside the larynx are the *vocal cords*. These are two folds of tissue extending across the opening of the larynx. They are attached to the *vocal muscles* by ligaments. These muscles produce a tension on the vocal cords, so that the air passing over them causes them to vibrate and produce a noise. When the muscles are fully relaxed, the larynx is fully opened and there is no noise.

(e) *Lungs.* The trachea divides into a pair of *bronchi* which further divide into the finer *bronchioles*. The ultimate branches of the bronchioles which are nearly as fine as capillaries terminate in the lung tissue proper. This consists of the thin-walled, lobed *alveoli*. The alveoli are provided with blood capillaries, and it is through the boundary formed by the walls of the alveoli and the blood capillary walls that gaseous exchange takes place between the external atmosphere and the blood (*see* Fig. 47).

(*f*) *The thoracic cavity.* The lungs are completely enclosed in the thoracic cavity. The base of the thoracic cavity is

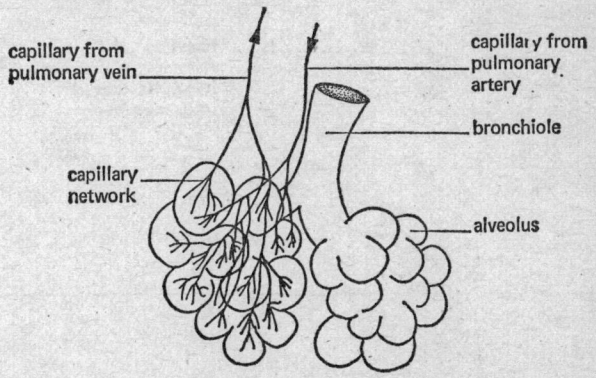

capillary from
pulmonary vein

capillary from
pulmonary
artery

bronchiole

capillary
network

alveolus

FIG. 47. Fine structure of lung.

a curved sheet of muscle—the *diaphragm*. Laterally the cavity is enclosed by the *ribs* and *intercostal muscles*. Between the lungs, and partially covered by them, is the *heart*.

(*g*) *Pleural membranes.* These fine membranes cover the surface of the lungs and line the inside of the thoracic cavity. They produce a fluid which lubricates the lung surface and pleural cavity allowing them to slip freely over each other during breathing.

4. Mechanism of breathing. The intake of air is called *inspiration*, and the expulsion of air, *expiration* (*see* Fig. 48). The normal rate of breathing is 15 times per minute, but this increases during activity, or when frightened, etc., and decreases during sleep.

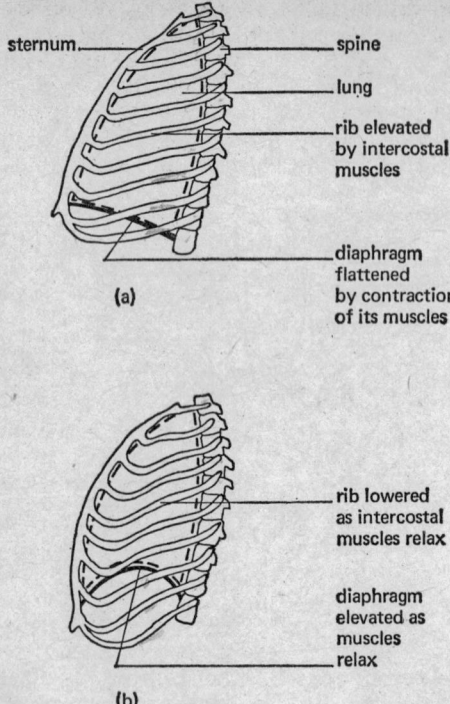

sternum — spine

— lung

— rib elevated
by intercostal
muscles

— diaphragm
flattened
by contraction
of its muscles

(a)

rib lowered
as intercostal
muscles relax

diaphragm
elevated as
muscles
relax

(b)

FIG. 48. Mechanism of breathing: (a) inspiration; (b) expiration.

(a) *Inspiration.* The lungs are enclosed in the thoracic
cavity, which is air-tight, so that their only opening to the
outside is through the trachea.

During inspiration:

(i) The external intercostal muscles contract, moving
the ribs outwards and upwards. This increases the diameter
of the chest.

(ii) The muscles of the diaphragm contract, flattening it
and increasing the depth of the thoracic cavity.

(iii) These muscular movements increase the volume of

VI. HOMEOSTATIC MECHANISMS

the thoracic cavity, and hence reduce the pressure within it. When the pressure is below that of the atmosphere, the air from the outside is forced into it through its own pressure. N.B. There is *no* active sucking mechanism drawing air into the lung.

(*b*) *Expiration*. This is largely a passive process. The external intercostal muscles relax, and the ribs drop under their own weight. The muscles of the diaphragm relax, and the lungs contract under their own elasticity expelling the air. More air can be forced from the lungs, by the contraction of the internal intercostal muscles and the abdominal muscles. This latter forces the organs of the abdomen against the diaphragm.

5. Lung capacity. The human lung contains about 4·5 l of air. This is called the *vital capacity*. During gentle breathing, *e.g.* when resting, only some 500 cm³ of this air is exchanged. This is called *tidal air*. If inhalation is forced, a further 1·5 l of air can pass into the lungs. This is *complemental air*. It is possible to force a further 1·5 l of air out of the lungs after the tidal air is expelled. This is *supplemental air*. Even after forced exhalation, the lungs are never empty. There is always about 1 l of *residual air* left in them.

6. Composition of the air. Table I shows the change in the composition of the air during breathing.

TABLE I. CHANGE IN THE COMPOSITION OF THE
AIR DURING BREATHING.

	Inspired	Expired
Oxygen	21%	16%
Carbon dioxide	0·03%	4%
Nitrogen	78%	78%
Water vapour	variable	saturated

7. Control of breathing. The movement of the intercostal muscles and diaphragm muscles is controlled by nerves from the "respiratory centre" of the brain. This is in the *medulla*. It is affected by the concentration of carbon dioxide in the blood. A high carbon dioxide concentration stimulates the respiratory centre to produce impulses which accelerate the expansion and contraction of the breathing muscles. This increases the rate of ventilation of the lungs, which will result in a decrease in the blood carbon dioxide. The stimulus is thus

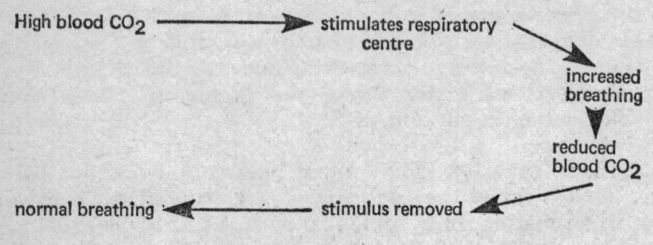

A homeostatic mechanism

removed from the "respiratory centre" which ceases to initiate the increased activity of the respiratory muscles. Thus, increased muscular activity (which will cause an increase in blood carbon dioxide) will indirectly cause an increase in the rate of breathing.

This is a good example of a homeostatic mechanism and illustrates the feed-back principle as applied to biological systems.

8. Diseases of the lungs. (Tuberculosis and lung cancer will be discussed in X.)

(*a*) *Pneumonia.* There are several forms of pneumonia but they all involve inflammation of the lungs. The disease is caused by a bacterium, and is characterised by a very high temperature. It affects people of all ages, but particularly the very young and old. Pneumonia is often a complication of another illness, especially in old people whose resistance has been lowered.

(*b*) *Bronchitis.* This is an inflammation of the trachea and bronchi. Constantly breathing in polluted air, and cigarette smoking are the commonest causes of bronchitis, although a damp climate may be contributory. Mucus is produced by the walls of the inflamed passages, causing a frequent cough. Blocking of the respiratory passages with mucus makes them inefficient, and breathing difficult. This reduces the available oxygen to the tissues, especially the heart muscles. Under exertion, the demand for oxygen by the muscles causes an increase in heart beat, and as the heart is starved of oxygen itself, heart failure can result.

(c) *Pleurisy*. This is an inflammation of the pleural membranes.

(d) *Asthma*. This is a nervous disease. The nerves of the bronchioles cause them to contract suddenly. Attacks may be initiated by excitement or nervous tension.

THE LIVER

9. Introduction. The liver is the largest organ in the body. It is dark red in colour, and lies immediately below the diaphragm overlapping the stomach on the right side. There are four lobes of varying sizes, and the whole organ is held in position by ligaments. On the underside of the liver lies the gall bladder, which stores bile.

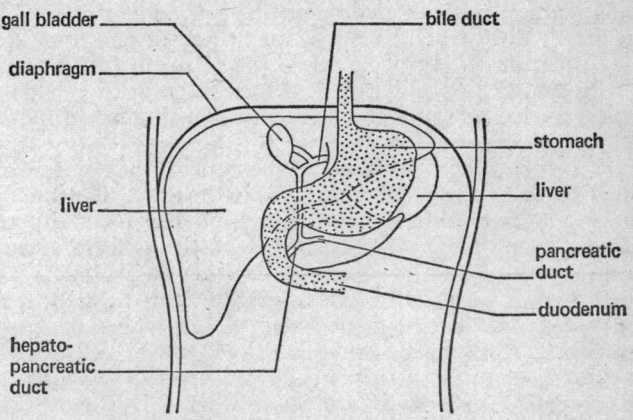

Fig. 49. Position of liver.

There are *three* blood vessels supplying the liver, the *hepatic artery, hepatic vein,* and *hepatic portal vein.*

The liver is the chemical control centre of the body, and is one of the most important organs in maintaining the constant composition of the body fluids (especially the blood) which form the immediate environment of the tissues (*see* Fig. 49).

10. Functions.

(*a*) *Regulations of blood sugar.* The blood normally contains about 0·1 per cent glucose. Glucose, amino acids and other food substances are brought to the liver from the ileum by the hepatic portal vein. Those food substances required for immediate use pass through the liver into the hepatic vein, and hence to the other tissues of the body. Any excess is converted by enzymes in the liver to the carbohydrate *glycogen.* Conversely, if the blood contains less than 0·1 per cent glucose, the glycogen in the liver is converted, by enzymes, into glucose. The regulation of the blood sugar is under the control of the hormone insulin.

(*b*) *Deamination of amino acids.* The body cannot store amino acids or protein. Those that are not immediately required by the tissues are deaminated in the liver. Amino acids contain carbon, hydrogen, oxygen and nitrogen. In the liver, the carbon, oxygen, and some of the hydrogen are converted to glycogen, while the nitrogen (in the form of the amino group NH_2) is converted to *urea.* This is passed back into the blood stream and is eliminated by the kidneys.

(*c*) *Detoxication.* Poisonous substances that enter the blood stream are rendered harmless in the liver. The innocuous products are then discharged from the blood by the kidneys. One of the chief sources of these harmful substances is the action of intestinal bacteria on proteins.

(*d*) *Formation of bile.* Bile is manufactured partly from the breakdown products of haemoglobin which has been released during the destruction of red blood cells in the spleen. The result of this breakdown are the green-brown bile pigments *bilirubin* and *biliviridin.* The liver also produces the alkaline bile salts. Bile is stored in the gall bladder from whence it passes down the bile duct to the duodenum. Here it helps to produce the alkaline conditions necessary for the action of the digestive enzymes, and emulsifies fats.

(e) *Storage of iron.* Iron is one of the products of the breakdown of haemoglobin. This, and iron from the food, is stored in the liver.

(f) *Production of antibodies.*

(g) *Production of fibrinogen.* This is a soluble protein found in the blood plasma. It is important in the formation of blood clots.

(h) *Production of antiprothrombin* (*heparin*). This is an enzyme which prevents the blood clotting, except at wounds.

(i) *Storage of vitamins.* Vitamins A, B complex, and D are stored in the liver.

(j) *Storage of blood.* The liver contains many blood vessels in which blood can be stored.

(k) *Production of heat.* Many of the reactions mentioned above involve the release of heat which is distributed to the rest of the body through the circulatory system.

THE KIDNEYS AND EXCRETION

11. Introduction. The kidneys are a pair of bean-shaped organs lying high in the abdomen and tucked under the diaphragm. The left kidney is slightly higher than the right. Each kidney is about 10 cm long and 5 cm wide. They are attached to the muscles of the back, and frequently there are deposits of fat around them.

Blood enters the kidneys through the renal arteries, and leaves them through the renal veins.

From the kidneys run the *ureters* which carry *urine* from the kidneys to the bladder (*see* Fig. 50).

12. Structure of the kidney. If the kidney is cut longitudinally, it is seen to contain three distinct layers. The outer darker region is the cortex, inside which is the lighter coloured *medulla* which projects as a number of cones (*pyramids*) into a space (the *pelvis*) which extends into the ureter (*see* Fig. 51).

In the cortex, the renal artery forms small knot-like groups of capillaries, each of which is called a *glomerulus.* Around each glomerulus is a cup-shaped structure called the *Bowman's*

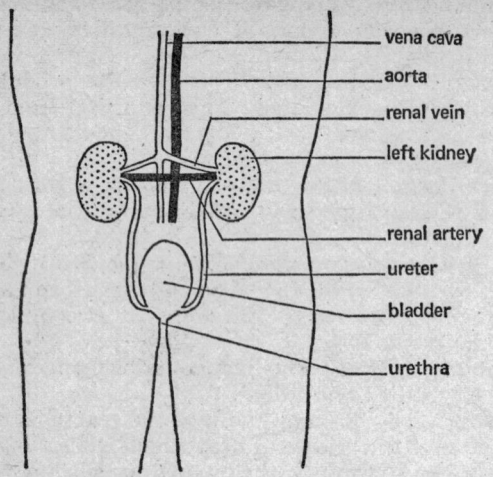

FIG. 50. Position of the kidney.

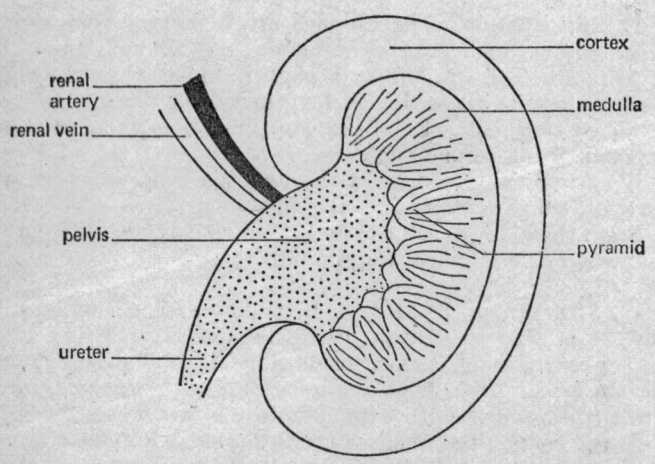

FIG. 51. Section through the kidney.

capsule. From the Bowman's capsule leads a long-looped *uriniferous tubule.* Several uriniferous tubules join to form a common *collecting duct* which opens into the pelvis through the

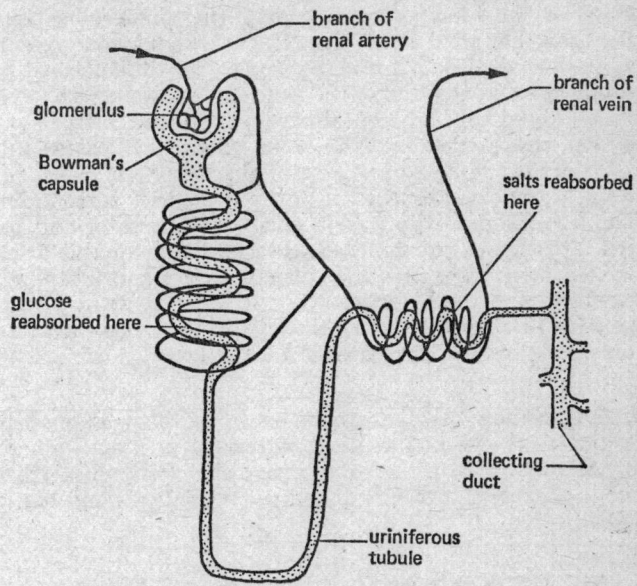

branch of
renal artery

branch of
renal vein

glomerulus

Bowman's
capsule

salts reabsorbed
here

glucose
reabsorbed here

collecting
duct

uriniferous
tubule

Fig. 52. Diagram to show relationship between blood vessels and other kidney ducts.

pyramid. Loops of the collecting ducts and uriniferous tubules pass into the medulla. Further branches of the renal artery form a network around the uriniferous tubule and ultimately fuse to form the renal vein (*see* Fig. 52).

13. Functions of the kidney.

(*a*) *Osmoregulation.* This is the regulation of the amount of water and dissolved substances in the blood plasma.

(*b*) *Excretion.* This is the removal of waste substances which are produced by the metabolism of the body cells.

14. Mechanism of controlling the composition of the blood plasma. The blood enters the kidney through the renal artery, and as it is forced into the fine capillaries of the glomerulus the pressure is increased. This pressure is reinforced by the diameter of the blood vessels leaving the glomerulus being smaller than that of those entering it. This high blood pressure forces the blood plasma through a molecular filter formed by the capillaries and the wall of the Bowman's capsule. The fluid which is filtered through contains water, glucose, nitrogenous waste, *e.g.* urea, and dissolved salts. This then passes into the uriniferous tubule where a lot of the water and salts, and all the glucose is re-absorbed into the capillaries surrounding it. The liquid left in the tubule now contains water and unwanted nitrogenous substances and salts. This liquid is *urine*. The urine then passes into the collecting duct and ultimately to the ureter and the bladder.

About 180 l of liquid passes through the tubules each day, but only 1·5 l of this is released from the body as urine.

15. Composition of urine. Urine contains about 95 per cent water and 4–5 per cent dissolved materials (urea and mineral salts). Abnormalites in the composition are often indicative of disease, *e.g.* excess glucose is diagnostic of diabetes and bile of jaundice.

16. The bladder and urination (micturition). The bladder is a muscular sac lying low in the abdomen. It opens to the exterior by the *urethra*. In the male, the urethra also serves as the duct through which sperm are expelled.

At the junction of the bladder and urethra is a ring of muscles which closes the urethra. When this ring of muscle relaxes, urine is expelled by the contraction of the muscles of the bladder wall and the muscles of the abdomen. This process of micturition is triggered off when the muscles of the bladder are stretched by the pressure of the urine inside it. (The bladder can hold some 400 cm^3 of urine.) The stretching of the muscles stimulates stretch-receptors in the bladder wall, initiating the events leading up to the expulsion of urine. Micturition is basically a reflex action but control of the urethral sphincter is learnt through training, so that by the time that a child is about 2 years old, micturition becomes a controlled reflex.

The inability to control micturition is called *enuresis*. It is not uncommon among children, particularly at night. Normally it can be prevented by reducing the amount of liquid drunk before going to bed. In some cases enuresis continues beyond childhood into adolescence. It can be a very embarrassing complaint which requires a great deal of sympathy and understanding from parents and friends. The cause is often emotional, so that scolding and ridicule are useless in helping to overcome the difficulty.

17. Other excretory systems.

(*a*) *The lungs*. Water and carbon dioxide are removed from the lungs. Both these are products of cell metabolism, and are therefore excretory products.

(*b*) *The skin*. Water, salts and a little urea are lost during sweating.

(*c*) *The intestine* contains bile pigments which have been formed in the liver, together with water and salts.

NOTE: The undigested remains of food which make up the bulk of the faeces are *not* excretory products as they have never taken part in cell metabolism.

TEMPERATURE REGULATION

18. Introduction. The body temperature of man is 36·8°C (98·4°F). This is kept constant by the activities of the body. Animals that keep their body temperatures constant (birds, and mammals) are *homoiothermic*. The body temperatures of other animals vary with that of their surroundings. They are *poikilothermic*. Animals that are homoiothermic are independent of their surroundings, maintaining their body activities uniformly throughout variations in the external temperature.

Except for the hottest climates, the air temperature is normally below the normal body temperature of a man. This means that the body is constantly losing heat, in the following ways:

(*a*) Three-quarters of the heat is lost by *convection* and *radiation*.

(b) *Evaporation* of sweat.

(c) *Heating* of the air in the lungs and evaporation from the lungs.

(d) *In urine and faeces*.

This heat loss has to be replaced. This is achieved by the oxidation of food during the respiration of the cells. Increased muscular activity increases the amount of oxidation in the muscles and therefore produces more heat.

The heat control mechanisms mentioned below are controlled by reflexes initiated by the heat control centre in the brain. This, in turn, is influenced by the blood temperature.

19. Prevention of and compensation for heat loss.

(a) *The metabolic rate* increases to produce more heat.

(b) *Sweat production* is decreased, reducing the loss of body heat by evaporation.

(c) *The blood vessels in the skin contract* (*vaso constriction*). This restricts the amount of blood coming near the body surface and hence reduces heat loss by convection and radiation.

(d) *Shivering* increases muscular activity which produces heat.

(e) *Raising the hair* by the contraction of the erector muscles in the skin. The raised hair traps a layer of still air against the skin surface. This still air insulates the surface against heat loss. As the human body has such a fine covering of hair, this method of heat retention is not particularly important. The contraction of the erector muscles only produces "goose pimples".

The above are reflex actions, but a person can make an active attempt to keep warm by one of the following methods:

(a) Eating more food of a high calorific value;

(b) Wearing clothes which are good insulators;

(c) Taking vigorous exercise;

(d) Drinking hot drinks.

20. Prevention and compensation for overheating.

(a) *The blood vessels in the skin dilate* (*vasodilation*). This allows more blood to come near the body surface through which heat is lost by convection and radiation.

(b) *Sweating*. The increased activity of the sweat glands produces a film of sweat on the surface of the skin. The latent heat required for its evaporation is taken from the body. This reduces the body temperature.

A person may attempt to keep cool by:

(a) Reducing muscular exertion;

(b) keeping in the shade;

(c) wearing light-coloured clothes which reflect the rays of the sun.

Drinking cool drinks may have little value in reducing heat loss but compensates for the loss of sweat. After copious sweating a little table salt should be added to the drinks to compensate for the loss of salt during sweating.

Eating ice cream and drinking sweet drinks actually increases the body heat output as both these contain high calorie foods.

21. Clothing.

(a) Clothing worn next to the skin should be made of *materials that will absorb sweat*. This makes them more comfortable to wear and allows the sweat to evaporate. In this respect natural fibres, *e.g.* cotton, linen and wool are probably better than man-made fibres, *e.g.* nylon, but recently man-made fibres have been greatly improved in this respect.

(b) *The looser the weave* the more comfortable will be the material because:

(i) It will absorb sweat more readily;

(ii) it will retain more air which will insulate against the rays of the sun and heat loss.

(c) *Clothes should fit well*. If they are too tight they can constrict the blood vessels in the skin. Tight garments, *e.g.* corsets, can constrict the organs of the body, especially those of the abdomen. Clothing that is too slack chafes the skin and is uncomfortable.

(d) *Clothes should be kept clean*. They absorb sweat from the body and pick up dust from the surroundings.

22. Young children and temperature control. Because of their high surface/volume ratio, young children are particularly vulnerable to changes in the external temperature.

Babies in particular are unable to take any avoiding action to alter their immediate environment, so extra care is necessary to see that they do not suffer from extremes of temperature.

Babies should not be left in prams in strong sun. The pram should always be shaded. Their skin should not be exposed to strong sun for any length of time. Light-weight nappies and cotton dresses or rompers are suitable wear, while older children playing outside should wear a brimmed hat.

There is no reason for keeping a baby indoors because the weather is cold, as long as he is well wrapped up in light-weight blankets, with a hat covering the ears. Direct winds, dampness and cold winds should be avoided. Never put a hot-water bottle in a pram with a baby, he may burn himself on it.

In bed light-weight blankets should be used. Young babies should not have a pillow; they may be suffocated by it. In very cold weather, background heating is needed in a child's bedroom. This can be either central heating, or a low-wattage *covered* electric heater.

Children's clothes should be well fitting, remembering that they grow out of them very quickly. They should be light-weight and made of a material that can be washed and dried quickly. Except for underwear, man-made fibres are ideal for children's clothes.

23. Footwear.

(a) *Footwear should fit well.* The shoe should be long enough to allow ample movement of the toes, and with growing children, room to grow without cramping. Shoes should be wide enough to allow maximum lateral expansion of the toes. Shoes with pointed toes should not be worn as they can lead to deformation of the toes, especially an enlargement of the joint of the big toe.

(b) *Heels.* These should be broad enough to be able to stand and walk comfortably. If they are too high, they throw the body forward, leading to a bad posture and tired back and leg muscles. High heels also force the toes into the toe of the shoe and so can cause cramping of the toes.

(c) *Material.* The feet sweat and so need free access to air. Therefore the material from which shoes are made should be porous. Wellington boots (made of rubber) are excellent for keeping the feet dry, but should not be worn for a long time. Various plastics are used to make shoes. These shoes are relatively cheap, but are not porous. Leather is the best material for making shoes, but it is expensive. It is porous and wears well. In hot weather, or in the house, light-weight shoes or sandals should be worn to allow free access of air to the feet.

PROGRESS TEST 6

1. What is meant by "homeostasis"? (1)
2. Through what organs does gaseous exchange take place? (2)
3. What is the function of the larynx? (3)
4. What is the normal rate of breathing of a man? (4)
5. What parts are played by the intercostal muscles and the diaphragm in breathing? (4)
6. Define the following terms: (a) vital capacity, (b) tidal air, (c) complemental air, (d) supplemental air, (e) residual air. (5)
7. What is the composition of inspired air? (6)
8. What is the composition of expired air? (6)
9. What part of the brain controls breathing? (7)
10. Name four diseases of the lungs. (8)
11. Name the blood vessels connected to the liver. (9)
12. How does the liver control the concentration of blood sugar? (10)
13. Outline the process of deamination. (10)
14. Give four other functions of the liver. (10)
15. What are the functions of (a) Bowman's capsules, (b) uriniferous tubules? (14)
16. Name three other excretory organs besides the kidneys. (17)
17. What is meant by the following terms: (a) homoiothermic, (b) poikilothermic? (18)
18. State four means by which the body loses heat. (18)
19. Give four ways by which heat loss is prevented or compensated for. (19)
20. Why are young children particularly vulnerable to changes in temperature? (22)
21. Give three characteristics necessary in good footwear. (23)

EXAMINATION QUESTIONS

1. Define excretion. What parts are played by: (a) the liver; (b) the lungs; and (c) the skin; in excretion?

2. Draw a section through the thorax to show the position of the lungs and related organs. How does gaseous exchange take place in the lungs?

3. Draw a longitudinal section through a kidney to show the main regions. Explain the process by which urine is formed by the kidney.

4. How is heat: (*a*) lost; and (*b*) produced by the body? Outline the methods by which the body temperature is kept constant.

5. If you were buying clothes and shoes for a young child for what particular points would you look to make sure that they would be comfortable?

CHAPTER VII

DEFENCE OF THE BODY

THE SKIN

1. Introduction. The skin makes up approximately one-sixth of the weight of the body. It covers the complete external surface and so acts as the body's first line of defence against desiccation and the invasion of disease-causing organisms. As well as playing an important part in the control of body temperature, the skin is a sense organ. It is the first organ through which a man comes in contact with his external environment, and contains sense organs of touch and temperature.

2. Structure and function. The skin is divided into two layers, the *epidermis* and *dermis*.

(a) *The epidermis* is made up of three layers:

(i) *The cornified layer* is on the outside. It is made up of flattened dead cells. This layer is the primary defence of the body against the external environment. It resists abrasion; it is water-proof; and protects against the invasion of disease causing organisms. As the skin is rubbed, cells of the cornified layer are worn away and replaced from the layer below. On the parts of the body where the rubbing is greatest, *e.g.* the soles of the feet and palms of the hands, the cornified layer is especially thick, and raised up into calluses of hard skin.

(ii) *The granular layer* lies below the cornified layer. Its cells are pushed towards the outside forming the cornified layer. As this happens the cells become thickened and flattened, and ultimately die.

(iii) *The Malpighian layer* lies beneath the granular layer. The cells are actively dividing, the new cells being pushed up into the granular layer, so that the replacement of the cells of the cornified layer depends ultimately on the division of the cells of the Malpighian layer. This layer contains varying amounts of the dark pigment *melanin*, the amount of which determines the colour of the skin.

(b) *The dermis* is thicker than the epidermis. It is a composite organ made-up largely of connective tissues, but containing a number of other structures:

(i) *Sweat glands* are coiled tubes opening on the surface of the skin through pores. Their main function is to control body temperature (*see* VI). If the body temperature rises 0·2–0·5°C above normal, the sweat glands produce sweat. In a hot climate the amount of sweat produced can be up to 1 l/hr. Sweat is produced by the active absorption of water and dissolved salts from the surrounding tissues. The dissolved substances include sodium chloride and urea. Both water and sodium chloride have to be replaced by drinking water.

(ii) *Capillaries* run throughout the dermis, supplying food materials and oxygen to the tissues and removing excretory products. They also play a part in temperature control (*see* VI).

(iii) *Hair follicles* are outgrowths of the Malpighian layer. The cells at the base of the follicles divide rapidly, and become impregnated with *keratin* (the same substance that thickens the cells of the cornified layer). These dead cells are pushed upwards to form the hair, growth of which continues for some 3–4 years, after which time the hair falls out and a new period of growth begins.

(iv) *Erector muscle* is attached to the hair follicle. When the body is cold, or the person is frightened, the erector muscles contract lifting the hair vertically (*see* VI, **19**).

(v) *Sebaceous glands* produce an oily fluid. This is discharged into the hair follicle and keeps the skin and hair supple. It also helps to water-proof the skin and hair, and is mildly antiseptic.

(vi) *Sense organs* of touch, pressure, and temperature are found in the dermis.

(vii) *Subcutaneous fat* is stored in cells immediately below the dermis. This is a food reserve and an insulating layer to prevent heat loss (*see* Fig. 53).

3. Care of the skin.

(a) The surface of the skin is always moist with sweat and the secretion of the sebaceous gland (sebum). Because of this it traps dust and bacteria and the dead cells that have been rubbed off the surface of the skin. This film can block the pores in the skin and form black heads. These

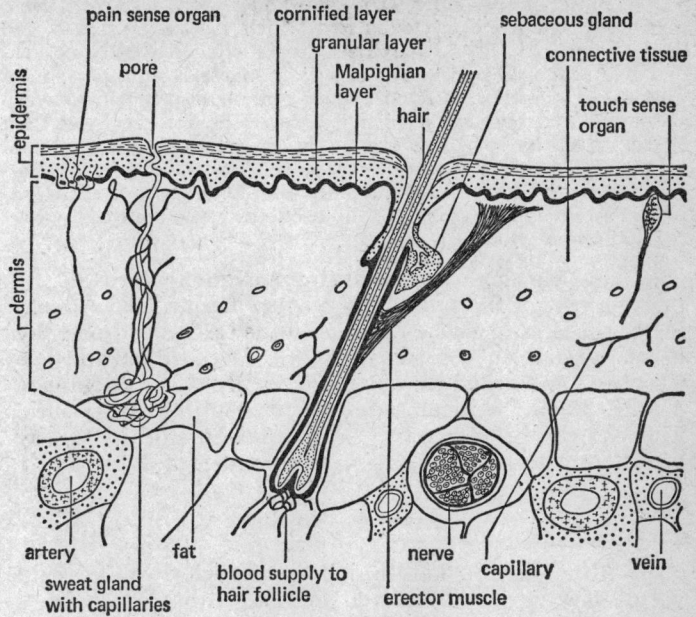

pain sense organ   cornified layer   sebaceous gland

granular layer

pore   Malpighian   connective tissue
           layer

hair   touch sense
            organ

epidermis

dermis

artery   fat   nerve   capillary   vein

sweat gland   blood supply to   erector muscle
with capillaries   hair follicle

FIG. 53. Section through the skin.

may turn septic and ultimately develop into a permanently spotty skin (acne). Adolescents are particularly prone to these skin disorders which accompany the changes in glandular activity.

(b) The skin (especially that of babies) can become rubbed and chapped. The former by clothing (babies are particularly prone to rubbing by damp napkins) and the latter by rough weather.

(c) Skin care involves:

(i) A good circulation which ensures a sufficient supply of oxygen to the cells and the removal of excretory products;

(ii) a balanced diet;

(iii) exercise;

(iv) fresh air;

(v) regular washing with soap and water. This removes the accumulation on the surface which blocks the pores. It also prevents the development of an offensive body odour which is caused by the by-products of bacteria multiplying in the skin secretion. A good quality (not necessarily expensive) soap should be used as poor quality soaps are often caustic. Detergents may "dry" the skin by removing too much sebum. Some chemicals used industrially can cause a dermatitis. Various forms of make-up should be used sparingly, and removed at night. If they are left on the skin the pores can become blocked.

(d) Sun-burning can result from over-exposure to the burning rays of the sun. This can vary from a mild reddening of the skin to severe damage to the tissues and possibly death. (Such an extreme is rare.) Dark-skinned people adapt to exposure to the sun more readily than fair-skinned people. Babies' skin is particularly susceptible to sun-burn. The skin can be adapted to tolerate the rays of the sun by gradual exposure, increasing over a number of days.

4. Care of the hair.

(a) Although the hair has little biological value (bald people survive equally as well as those with a fine head of hair), it has great social and psychological significance. So great is this that women (in particular) are willing to spend a great deal of time and money in hairdressers, and the production and sale of hair shampoos, conditioners, etc. is a large commercial project. (Look at the number of advertisements for hair shampoos on television and in women's magazines.)

If the hair is not cared for it soon becomes greasy and dirty with the accumulation of sebum deposited on it. This makes the hair unattractive to look at. If the hair is not attended to for a long period, it may well begin to harbour lice.

(b) The best ways to care for the hair are:

(i) *Regular washing and combing.* Apart from removing tangles and dust in the hair, this give it an attractive sheen and removes the eggs of parasites which may be lodged in it. Brushing, in particular, massages the scalp, stimulating the blood supply to the hair follicles. This encourages the growth of the hair.

(*ii*) *The hair should be washed* at least once a week, more if one lives or works in a city or factory, or if the hair tends to be greasy. A good quality shampoo should be used rather than soap, especially if the water is hard. Soap tends to form a scum which becomes deposited on the hair.

Regular washing, particularly with a shampoo containing specialised ingredients, discourages the development of dandruff. This complaint is not serious, but causes the skin of the scalp to flake off. It leads to a weak growth of the hair and to unsightly pieces of skin on the shoulders of clothes.

5. Diseases and parasites of the skin and hair.

Most skin complaints are contagious and are the direct result of a lack of personal cleanliness, or contact with people carrying the parasites. Infestation with animal parasites, *e.g.* lice and fleas, is frequently associated with poor and overcrowded living conditions, but infestation is equally possible among crowds, *e.g.* in cinemas (*see* Fig. 54).

(*a*) *Lice*. These are wingless insects. Their body is flattened and the adults maintain their hold on the body by gripping with their clawed feet. They feed on blood which they suck from the host through specialised mouthparts. There are three species of louse, each inhabiting a particular part of the body:

(*i*) *Pediculus capitis* lives attached to the hairs of the head. The eggs (nits) are small and white, attached to the hairs by a sticky substance. They hatch out in about 6 days, the young reaching maturity in a further 10 days. Head lice spread rapidly among a group of people in close contact with each other, and are especially prevalent among children of school age.

(*ii*) *Pediculus pubis* (crab louse) is found on the hairs of the pubic region.

(*iii*) *Pediculus humanus* (*P. corporis*) is rarely found attached to the body. It lives and lays its eggs in the folds of clothes, leaving them only to feed. This louse can carry relapsing fever and typhus. As the louse needs human blood to survive, storing infested clothing for 11–12 days ensures that any adults are killed. Eggs can survive for over a month, and it is safer to disinfest clothes by treating with steam.

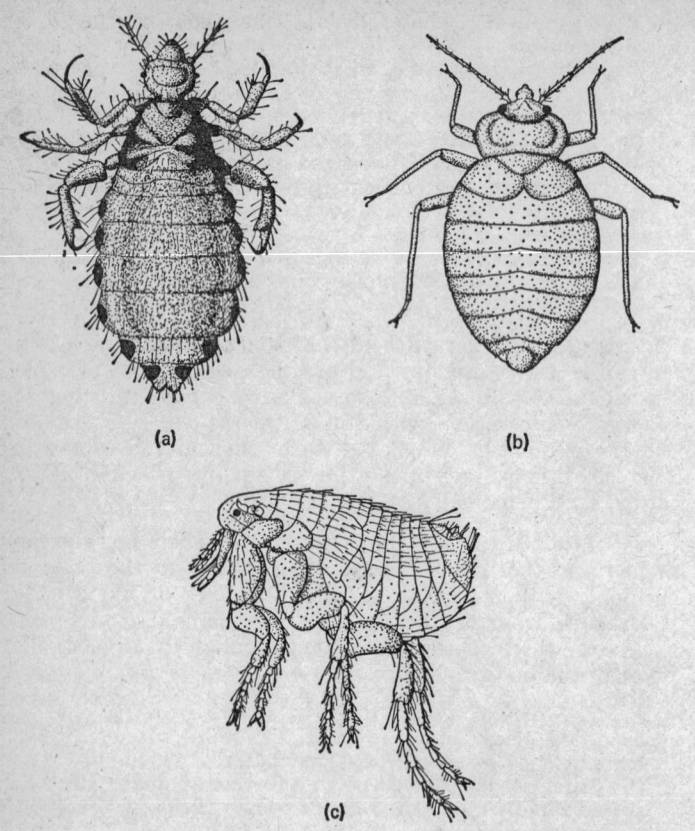

FIG. 54. Parasites of the skin and hair: (*a*) Louse (*Pediculus*);
(*b*) Bed bug (*Cimex*); (*c*) Flea (*Pulex irritans*).

Lice can be removed by treatment with insecticidal powders.

(*b*) *Fleas.* The human flea (*Pulex irritans*) is wingless, with the body compressed from side-to-side. The back legs are greatly enlarged for jumping and the feet are hooked to hold on to the hairs. They feed on human blood. Apart from the irritation caused by the bites and the possibility of the bites becoming infected, fleas are vectors of disease. The human flea can transmit bubonic plague from man to man, while the rat flea *Xenopsylla cheopsis* transmits plague from rats to man. Fleas die quickly if deprived of food. They can be eradicated by using an insecticidal powder.

(c) *Bugs*. The bed bug (*Cimex lectularis*) is wingless. It normally lives in cracks and crevasses in walls, floors, etc. (therefore it is associated with poor living conditions), coming out at night to feed on the blood of a human host. The bites cause great irritation and may become septic. Bugs may also transmit diseases. They can survive in buildings up to 9 months after their last blood meal, and are best controlled by fumigating the building.

(d) *Mites*. These small animals have eight legs and are related to the spider. One mite, *Acarus scabiei* causes *scabies*. This is the result of the female burrowing under the skin to lay her eggs. The male lives on the surface of the skin. The main areas of attack are between the fingers and toes. An intense irritation is caused. Scabies can be treated with benzyl benzoate or sulphur ointment; the latter should be used with care as there is a possibility of causing dermatitis.

(e) *Boils*. Boils are caused by an infection, usually of a hair follicle by a *Staphylococcus* (a bacterium). They are usually formed on the back of the neck where it is rubbed by the collar, but can occur anywhere where the skin is damaged. Invasion by the bacterium initiates a local defence reaction, leading to a swelling. The swelling may subside, but frequently develops into a head from which a pus of dead bacteria and white blood cells is discharged. A person is usually more susceptible to boils:

(i) If the diet is not balanced;
(ii) if they are not in good health;
(iii) during adolescence.

Treatment with an antiseptic solution of cream is usually sufficient, but in more serious cases antibiotics have to be used.

(f) *Acne*. This is caused by a bacillus and is instigated

by the blocking of the pores with dirt. Primarily it is a disease of adolescence and is particularly distressing as it occurs on the face and shoulders. As the sebaceous gland becomes inflamed a pimple (blackhead) is formed. A well-balanced diet and exercise are the best long-term cures, while antiseptic creams help in the short term. Covering the spots with cream or face powder does not help as they only block the pores further. Acne clears up after puberty.

(g) *Eczema*. This is a drying and reddening of the skin, which becomes inflamed. It is probably an allergy to particular proteins, so that woollen clothes should not be worn next to the skin. The allergy is more likely to occur in members of the same family. The disease is not contagious.

(h) *Ringworm* (see X).

(i) *Dermatitis*. This is a response of the skin to various chemicals. Tar products are the chief offenders, but many industrial chemicals are potential causes. The skin becomes reddened and flakes off. It can be avoided by wearing protective clothing, e.g. rubber gloves.

MUCOUS MEMBRANES

6. Introduction. As the external surface of the body is protected by skin, so the internal surfaces are protected by mucous membranes. Cells of the membrane are glandular, producing a slimy mucus. This lubricates particles, and prevents them from damaging the delicate membrane. Mucus-producing cells are found primarily in the linings of the intestine, nasal cavity, and respiratory passages.

7. The intestine. Production of mucus in the intestine, particularly the upper parts, lubricates the bolus of food. Mucus is a constituent of saliva which moistens and lubricates the food entering the mouth.

8. The nasal cavity. Mucus produced in the nasal cavity traps dust and bacteria entering the nose, and prevents their further passage into the respiratory system. Excessive stimulation of the membranes, e.g. by a dusty atmosphere or fumes (*e.g.* tobacco smoke or petrol fumes), leads to the production of large amounts of mucus and difficulty in breathing. This is called *catarrh*. A similar condition develops if the membranes become infected. The discomfort of the swollen membranes can be relieved by using drops or tablets containing decongestants. These may relieve the symptoms, but do not remove the cause. Severe or prolonged inflammation of the mucus membranes of the nasal cavity may lead to a blocking and infection of the sinuses in the bones of the skull, especially around the eye orbits (*synocytis*) which is very unpleasant and causes severe headaches.

9. The respiratory tract. Mechanical irritation of the mucosa, *e.g.* by dust, leads to an excess production of mucus which is coughed up by a reflex contraction of the muscles of the thorax and abdomen. Chemical irritation, *e.g.* by tobacco smoke, has the same effect giving a condition referred to as a "smoker's cough". Infection, often following excessive mechanical or chemical irritation, causes *bronchitis*.

DEFENCE MECHANISMS IN THE BLOOD

10. Blood clotting. The clotting of blood on the surface of a wound prevents loss of blood and the entry of disease organisms into the body. But the blood, which is usually free-flowing, must only clot when the blood vessels are damaged. The mechanism is as follows:

(*a*) The blood contains, dissolved in it, a protein called *fibrinogen* and a proenzyme *prothrombin*. It also contains an enzyme *antiprothrombin* (*heparin*) which prevents the prothrombin from being converted to thrombin.

(*b*) When a blood vessel is damaged, the cells lining it and the white cells produce *thromboplastin* which inhibits the action of antiprothrombin. Thus the prothrombin can be converted to *thrombin* which in turn changes the soluble fibrinogen to *fibrin*. Stages in the formation of fibrin require the presence of calcium ions and vitamin K.

(*c*) From the edges of the wound new skin cells grow across it beneath the clot, which, when the new skin is formed, is sloughed off as a scab.

Clotting of the blood in a blood vessel is a *thrombosis*. If such a clot occurs in a blood vessel of the brain, it causes a *stroke*. The blood supply (and hence the oxygen supply) is cut off from a section of the brain which ceases to function properly. This can lead to temporary, or permanent paralysis of part of the body. A *coronary thrombosis* is a blocking of the coronary artery which supplies the heart with blood. This can cause death.

11. Phagocytosis. This is a process whereby the white blood corpuscles (*phagocytes*) engulf bacteria that have entered the blood system, through a wound. If bacteria enter a wound, the phagocytes migrate to the site where they engulf and digest the bacteria. During this process the phagocytes themselves can die, and these together with the bacteria and dead tissues form *pus*. The formation of pus, together with the blood clot, prevent blood circulating through the area localising the infection. Any bacteria that escape from the infected area are destroyed in the lymph nodes, liver and spleen.

12. Immunity.

(*a*) The symptoms of many diseases are caused by the presence of foreign protein in the blood. This may be:

    (*i*) Bacteria;
    (*ii*) toxins produced by bacteria;
    (*iii*) viruses;
    (*iv*) breakdown products of infected tissues.

(*b*) The body reacts to the presence of these foreign proteins by producing *antibodies*. The protein stimulating the production of antibodies are called *antigens*. Antibodies are produced in the liver, spleen and lymphatic nodes. They cause bacteria to clump together (*agglutinate*) making them more readily ingested by the phagocytes. Special forms of antibodies called *antitoxins* are produced by the platelets. These denature bacterial toxins.

(*c*) Once the antibodies are produced, they remain in the blood for some time after the symptoms of the disease have disappeared, and give immunity to a further attack. The length of time that the immunity will last varies for different diseases. It may be a few months, or may be years.

13. Types of immunity.

(a) *Naturally acquired immunity* is present in a person that comes in contact with a disease organism, but shows no symptoms of the disease. This may be of four types:

(*i*) When a person lives in an area where a disease is always present (*endemic*) he will receive doses of the antigen sufficient to produce the antibody, but not sufficient to give marked symptoms.

(*ii*) After showing symptoms of the disease, a person has the antibodies in the blood which will prevent further symptoms being produced when a second dose of antigen is introduced.

(*iii*) Inherited immunity (this is a poor name as it is not inherited genetically) is acquired by babies from the mother through the placenta or from the milk. The antibodies are present in the mother's blood. This is one advantage of breast feeding rather than bottle feeding a baby. This immunity lasts only a few months by which time the baby can produce its own antibodies.

(*iv*) A person in a good state of health will not show symptoms of a disease as readily as one who is unwell, particularly through under-nourishment.

(b) *Artificial immunity* is produced by clinical treatment of a patient. There are two types:

(*i*) *Passive immunity*—produced by injecting a serum from another animal which has had a mild attack of the disease. That is, the antibodies are produced in the blood of another animal, and transferred to the human being where they can function normally. This type of immunity is short-lived.

(*ii*) *Active immunity*—the introduction of the antigen into the blood stream, which induces the production of the antibody. This process is called *vaccination*. There are three types:

I. inoculation with the live but attenuated (mild) strain of the organism, *e.g.* smallpox;

II. inoculation with the dead organism, *e.g.* whooping cough;

III. inoculation with a modified (harmless) form of the toxin.

14. Blood groups. During a blood transfusion the blood of one person (donor) is passed into the blood stream of another (recipient). Blood, being a foreign protein, may induce the production of antibodies when introduced into the blood of the recipient. When the antibodies are produced they cause the red blood cells to clump together (*agglutinate*).

There are two factors present in the red blood cells which can cause this agglutination. They are called A and B. Hence there are four *blood groups*—O (no factor present), A, B, and AB (both factors present).

If the blood contains a factor, it will not produce the antibody against it. Also, if it does not contain the factor, it will produce the antibody against it. Therefore:

(a) Blood group O produces antibodies *a* and *b*;
(b) blood group A produces antibody *b*;
(c) blood group B produces antibody *a*;
(d) blood group AB produces no antibodies.

Therefore a person of blood group O can only receive blood from blood group O, but can be used as a donor for any group, Similarly blood group AB can receive blood from any donor, but can only donate to another AB person. The complete set of possibilities is set out in Table II below:

TABLE II. BLOOD GROUPS OF DONORS AND RECIPIENTS.

| | | *Recipients* | | | |
		O	A	B	AB
Donors	O	yes	yes	yes	yes
	A	no	yes	no	yes
	B	no	no	yes	yes
	AB	no	no	no	yes

There is another form of blood group called the *Rhesus factor*, which occurs in Rhesus monkeys as well as human beings. This factor is inherited. The baby and mother are separate individuals genetically and otherwise so that it is possible for a mother to have a different blood group from that of her baby. This has two consequences:

(a) It is possible for the baby's blood to seep across the placenta and so cause the formation of antibodies against the baby's blood in the mother's blood. If these then enter the blood stream of the baby, they can agglutinate the baby's blood. This is an infrequent occurrence.

(b) A woman of child-bearing age should not be given blood (during a transfusion) of a Rhesus factor different from her own because antibodies will be formed which could destroy the blood of her child and kill it. The antibodies are long-lasting so that the woman need not be pregnant at the time of the tranfusion for this to take place.

PROGRESS TEST 7

1. What are the two main layers in the skin? (2)
2. What part does the skin play in the defence of the body? (2)
3. Give five ways in which to care for the skin. (3)
4. Give two ways in which to care for the hair. (4)
5. Name three parasites of the hair and skin. (5)
6. What organisms cause boils? (5)
7. Name three organs which are lined by mucous membranes. (6)
8. What is the main function of mucous membranes? (6–8)
9. Explain how the blood clots. (10)
10. Name four kinds of foreign protein which can cause diseases. (12)
11. What are the two types of immunity? (13)
12. Why is it important to know the blood groups of both the donor and the recipient before a blood transfusion is performed? (14)

EXAMINATION QUESTIONS

1. *Briefly* outline the life cycles of three skin parasites. How can attack by these parasites be controlled?
2. Draw a diagram of a section through the skin. How can the skin be kept healthy by everyday care?

3. Outline the methods by which the blood protects the body from infection.

4. Explain what is meant by a "blood group". Why is it important for a person to know his blood group?

CHAPTER VIII

HEREDITY AND REPRODUCTION

HEREDITY

1. Introduction. A number of plants and animals reproduce *vegetatively* or *asexually*. That is, a piece of the parent organism develops into a new individual. This can be a very efficient method of reproduction, but has the disadvantage that the individuals produced are identical to the parent, which means that they are unsuited to any change in the environment. A large number of plants and animals reproduce sexually and asexually which gives the reliability and rapidity associated with asexual methods, but allows for variation which is provided by the sexual method. Man and other mammals reproduce only sexually.

2. Sexual reproduction. Sexual reproduction is the fusion of *gametes* to produce a *zygote* which develops into a new individual. As the gametes are cells, the zygote will contain chromosomes from each parent (*see* I), and as is explained later can by this means inherit characteristics from each parent. The sexual method of reproduction has three disadvantages:

(*a*) When the gametes fuse together, there is a doubling of the number of chromosomes. To keep the number of chromosomes constant, at some stage in the life cycle there has to be a special cell division, during which the number of chromosomes is halved. This division is called *meiosis*. In human beings it takes place during the formation of the eggs and sperm.

(*b*) The eggs and sperm have to fuse together, so that there is always a chance that they will not meet. The chance of their meeting is greatly increased by the production of large numbers of gametes, *e.g.* by fish, when fertilisation takes place outside the body of the female, but this is a very wasteful method. Fertilisation in mammals takes place inside the body of the female. This increases the chance of the gametes meeting and reduces the number of gametes necessary.

(c) After fertilisation, the young embryo is very vulnerable to changes in the environment, and to predators. It also requires feeding until it can feed itself. Some animals, e.g. fish, produce a large number of eggs, which compensates for the number of embryos which are destroyed; others, like birds, protect the young in a shelled egg which is well provided with food; while the mammalian embryo develops within the female, where it is protected and fed through the mother's blood stream. Even after it is born, there is a high degree of parental care.

3. Meiosis (reduction division).

This form of cell division takes place during gamete formation, resulting in the halving of the number of chromosomes. The stages of division are similar to those of mitosis (see I).

(a) Each cell of the human body contains twenty-three pairs of chromosomes (*homologous chromosomes*). The members of each pair look alike.

(b) During the *first meiotic division* the nuclear membrane disappears and the homologous chromosomes come together in pairs. They split longitudinally into their respective *chromatids*, which remain attached at the *centromere*. The chromatids may become *crossed over*, the significance of which will be explained later. The pairs of chromosomes become arranged on the spindle to form the equatorial plate. Then the *complete chromosomes*, *i.e.* the pairs of attached chromatids, separate, moving to the opposite poles of the spindle. When they reach the poles, the spindle disappears. Thus, at the end of the first division, there are twenty-three chromosomes at each end of the cell, and each one of the chromosomes has its homologue in the other half of the cell.

(c) At the beginning of the *second meiotic division* two spindles form at opposite ends of the cell, and at right angles to the axis of the spindle formed during the first meiotic division. The chromosomes become arranged on these new

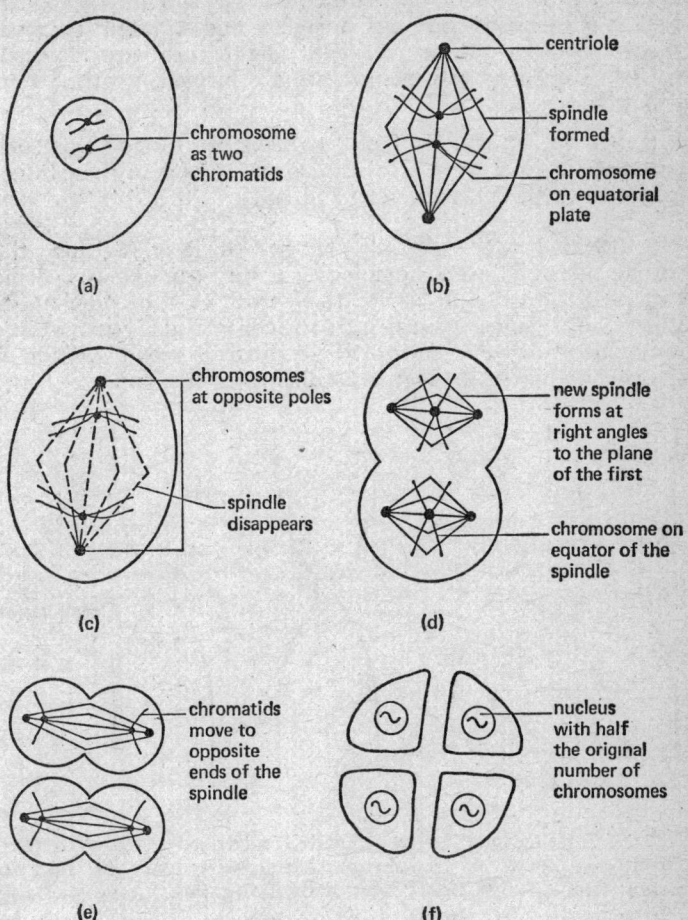

FIG. 55. Stages in meiosis.

spindles forming equatorial plates, and the *separate chromatids* now move to opposite poles. The spindles disappear. The chromatids become uncoiled and thicken to form chromosomes. New nuclear membranes are formed. Ultimately new cell membranes are formed, isolating four new cells.

Hence during meiosis four new cells are formed from one parent cell. Each new cell contains one of each of the pairs of homologous chromosomes (*see* Fig. 55).

4. Crossing over. During the first meiotic division, the chromatids of adjacent homologous chromosomes may become twisted together, and where they cross over become fused. When the chromosomes separate the chromatids break at these junctions (*chiasmata*) so that there can be a recombination of the genetic material. Figure 56 illustrates this point.

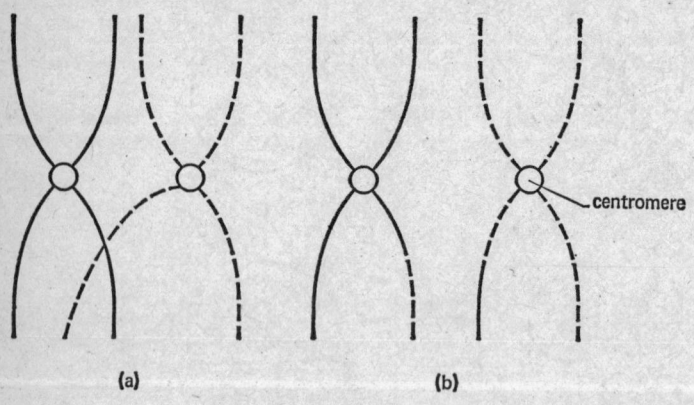

Fig. 56. Crossing over: (*a*) two chromosomes derived from different parents, with two of the chromatids crossed over; (*b*) the two chromosomes pulled apart with a breaking and rejoining of the crossed-over chromatids.

5. DNA and RNA. We saw in I that the nucleus is the control centre of the cell, and that the "controlling units" are the chromosomes. Each chromosome is a double strand of

DNA. DNA is *deoxyribosenucleic acid*. Each strand is a group of sugar (deoxyribose) groups alternating with phosphate groups, and the rungs are pairs of nitrogen bases (adenine, guanine, cytosine and thymine). Some groups of these nitrogen bases influence the activity of the cell and are called *genes*.

The genes function by coding the formation of RNA. RNA is *ribosenucleic acid* which is similar to DNA, except that the sugar is ribose, and uracil replaces thymine. The RNA is formed in the nucleus and moves through pores in the nuclear membrane to the ribosomes. On the ribosomes, the RNA in turn, codes the production of different proteins. Remember that enzymes are proteins, so that this mechanism can influence the activities of the cell. The gene function is summarised below:

DNA (gene) $\rightarrow$ code for $\rightarrow$ RNA $\rightarrow$ transfers $\rightarrow$ ribosomes
in nucleus   RNA   information $\downarrow$
stores   production
information   PROTEIN

When we mention genes in the next section, remember that these are small groups of nucleic acids, and that they can be considered as a row of points distributed along the length of the chromosomes. Referring to Fig. 57, it can now be seen that crossing over results in different combinations of genes being produced.

THE MECHANISM OF INHERITANCE

6. Mendelian inheritance.

(a) *Mendel's work*. The modern science of genetics (*i.e.* how characters are inherited) is based on the work of Gregor Mendel, who was a monk in the abbey of Altbrunn. He died in 1884 so had no knowledge of genes. His work was carried out on pea plants. He grew pea seeds, and from the plants he selected the tallest and the shortest and self-pollinated them. He then grew the resulting seeds, and from the plants he selected the tallest and the shortest and self-pollinated them. He grew the resulting seeds and repeated the process until the seeds from tall plants produced only tall plants, and seeds from short plants produced only short plants. Two *pure lines* had been established. Then he crossed tall plants with short plants and grew the seeds (the first filial or F_1 generation) which produced only tall plants. These F_1 tall plants were then cross-pollinated amongst themselves and the seeds (second filial or F_2 generation) produced both tall and short plants in the ratio of three tall to one short. This 3 : 1 ratio is called a *monohybrid ratio*. This meant that the "factor" (as Mendel called it) for shortness had not disappeared, but had been masked by the factor for tallness. The character that is masked (in this case shortness) is called the *recessive* while the one which shows is called *dominant*. These results led Mendel to propose his *first law of inheritance* which is:

Only one pair of factors affecting a particular character is present in a gamete.

(b) *Interpretation of Mendel's results.* The somatic cells of the pure tall plants would have a pair of homologous chromosomes each with a gene in the "height" position which would cause "tallness". When meiosis took place, each gamete would contain a gene for "tallness". Similarly, the "height" position on the short plant chromosomes would be occupied by a gene for "shortness". When the gametes from these pure lines fused, the zygote would have one gene for "tallness" and one for "shortness" but they would all appear tall because the "tallness" gene masked the effect of the "shortness" gene.

When meiosis occurred and gametes were formed by the F_1 generation, the "tallness" and "shortness" genes were again separated and could recombine at random. From the diagram opposite it will be evident how the 3 : 1 ratio appears.

(c) *Genotype and phenotype.* The *genotype* refers to the genes for a particular character that are present in the individual. If both genes are the same the individual is *homozygous* and if they are different he is *heterozygous*. The F_2 generation in the example above contains one homozygous tall individual, two heterozygous tall individuals and one homozygous short individual.

The *phenotype* describes the characteristic which actually shows in an individual. In the above example there are two phenotypes, namely "tall" and "short". Notice that two individuals can have the same phenotype, but different genotypes.

(d) *Incomplete dominance.* Eye colour in man illustrates this. There are a large number of eye colours, not only brown and blue. If both parents are homozygous for "brown" or "blue" their children will have brown or blue eyes, but if they are heterozygous "brown" some of the children will also be heterozygous and this may show in

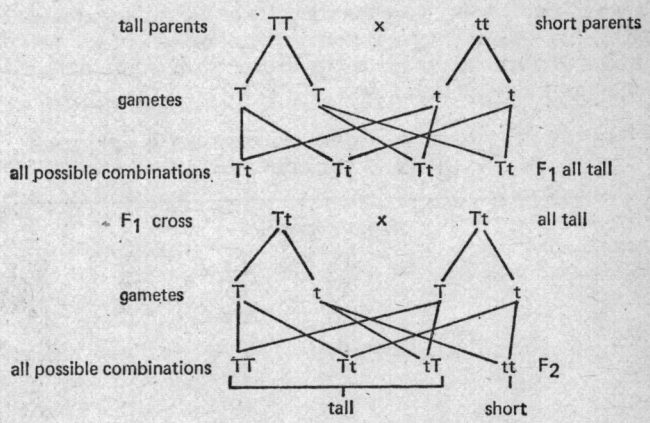

Mendel's First Law of Inheritance

their eyes being light brown or hazel. The "brown" gene is not completely dominant over the "blue" gene.

(e) *Other factors which confuse the issue:*

(i) Some characters may not be controlled by only one gene, *e.g.* "tallness" in humans is controlled by a large number of genes.

(ii) Genes may be prevented from showing their full potential, *e.g.* even if a person has all the genes necessary to grow tall, he will not be able to unless he has sufficient food.

(f) *Mendel's second law.* Mendel continued his experiments to determine what happened when two characters were inherited simultaneously. For this he crossed pure lines of tall peas with yellow seeds with short plants with green seeds. "Tall" is dominant to "short" and "yellow" is dominant to "green". The F_1 generation were all "tall yellow" plants, but the F_2 generation had "tall yellow", "tall green", "short yellow", and "short green" plants in the ratio of $9 : 3 : 3 : 1$. This is called the *dihybrid ratio*. This ratio can only be achieved if the factors for height and seed-colour are inherited independently of each other, *i.e.* there is an equal chance of height and colour genes being inherited together. If T is "tall", t is "short", Y is "yellow-seeded" and y is "green-seeded", the possible gametes are TY, Ty, tY, and ty. When these combine in all possible ways to form zygotes, there are sixteen combinations as shown in Table III below:

TABLE III. POSSIBLE COMBINATIONS OF TWO FACTORS INHERITED INDEPENDENTLY.

		Female gametes			
		TY	Ty	tY	ty
	TY	TY TY	TY Ty	TY tY	TY ty
male gametes	Ty	Ty TY	Ty Ty	Ty tY	Ty ty
	tY	tY TY	tY Ty	tY tY	tY ty
	ty	ty TY	ty Ty	ty tY	ty ty

Any plant containing the dominant T will be tall; plants with only t will be short. Plants with the dominant Y will be yellow-seeded and those with only y will be green-seeded. These results led Mendel to postulate his second law which states that:

Single characteristics associated with the parents can be inherited independently by the offspring.

If one takes each of the characteristics of height and seed-colour separately in Table III above, you will see that each is inherited on the $3 : 1$ basis. (If you are mathematically inclined you can show that $9 : 3 : 3 : 1$ is $(3 : 1) \times (3 : 1)$ and work out the ratio you would get considering three characteristics.)

It was mentioned earlier that more than one gene had to be present for some phenotypes to be produced. Draw another square like the one above, and using the dominant genes X and Y, work out the ratio if both dominant genes have to be present in the zygote to produce a phenotype. Use the same square work out the ratio if either gene X or gene Y only is necessary to produce a phenotype. (*Hint:* the gametes will be XY, Xy, xY, xy.)

7. Sex determination. Of the twenty-three pairs of homologous chromosomes present in the somatic cells of a human being, twenty-two of the pairs have members that look alike. These are called *autosomes*. The other pair are the *sex chromosomes*. In the female they are identical and are called X chromosomes, while the male has one X chromosome and a much shorter Y chromosome. When the gametes are produced, the eggs always have an X chromosome, while the sperm have X or Y chromosomes. Table IV below shows why the number of boys and girls born is approximately equal.

TABLE IV. SEX DETERMINATION

	Man		*Woman*	
Somatic cells	XY		XX	
gametes	X	Y	X	X
children	XX	XY	XY	XX
	boy	girl	girl	boy

8. Sex linkage. Some characters are determined by a recessive gene on the X chromosome and appear in the phenotype of the male. These characters are *sex linked*. Green-red colour blindness and haemophilia are two such conditions. Green-red colour blindness is rare among women and will only occur if the father is affected and the mother is heterozygous for the gene (a carrier).

The appearance of a sex-linked character only in the male is dependent on the Y chromosome not carrying the dominant gene necessary to mask the recessive. As the Y chromosome

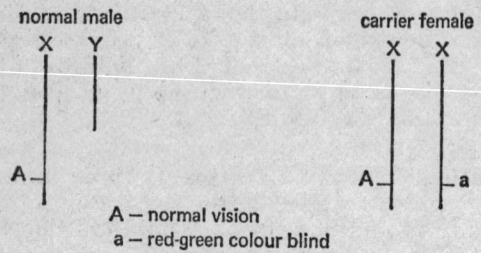

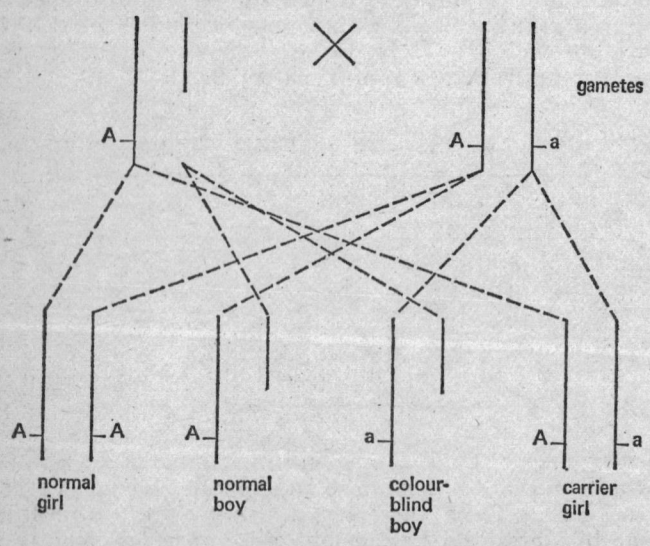

A — normal vision
a — red-green colour blind

Fig. 57. Sex linkage of colour blindness.

is shorter than the X it usually means that the gene is carried on that part of the X chromosome which has no corresponding part on the Y. Figure 57 will illustrate this point.

9. Mutation. Very occasionally, random, sudden changes take place in genes which give rise to new, abnormal characteristics. These changes are mutations. By these means new phenotypes can develop in a population. In experimental work on the fungus *Neurospora* and the fruit fly *Drosophila*, mutations have been produced by using X-rays, mustard gas, and various other chemicals. Both these organisms produce a large number of offspring in each generation, so the genetic effects can be studied.

REPRODUCTION

10. The male reproductive system.

(*a*) The paired *testes* are the male reproductive organs, which begin to produce sperm as a boy reaches puberty (10–14 years old). They are enclosed in the *scrotal sac* which extends outside the body cavity. Internally they consist of a large number of microscopic tubes—the *seminiferous tubules*—which are lined with cells that divide to form the *sperm*. Between the seminiferous tubules are blood vessels, connective tissue, and the *interstitial cells*. The interstitial cells produce the male hormone *testosterone*.

(*b*) The seminiferous tubules converge into fine *sperm tubules* which lie outside the testes. These in turn join to form the coiled *epididymis* (where the sperm are stored) which leads into the muscular *vas deferens*. This loops over the ureter, leading from the kidney to the bladder, to join the *urethra*. The urethra is a common duct through which urine and semen are passed. Before joining the urethra, the vas deferens passes through the *prostate gland* which surrounds the urethra, and immediately after the joining of the two ducts, the urethra passes through *Cowper's gland*. Both these glands add fluid to the sperm during copulation. This fluid contains nutrients and enzymes which stimulate the sperm into swimming. Until they are stimulated they are immobile.

(c) In the male the urethra is extended into an external *penis*. This organ consists of connective tissue through which blood vessels and blood spaces run. During copulation the blood spaces become gorged with blood, by more blood running into them than from them. This makes the penis

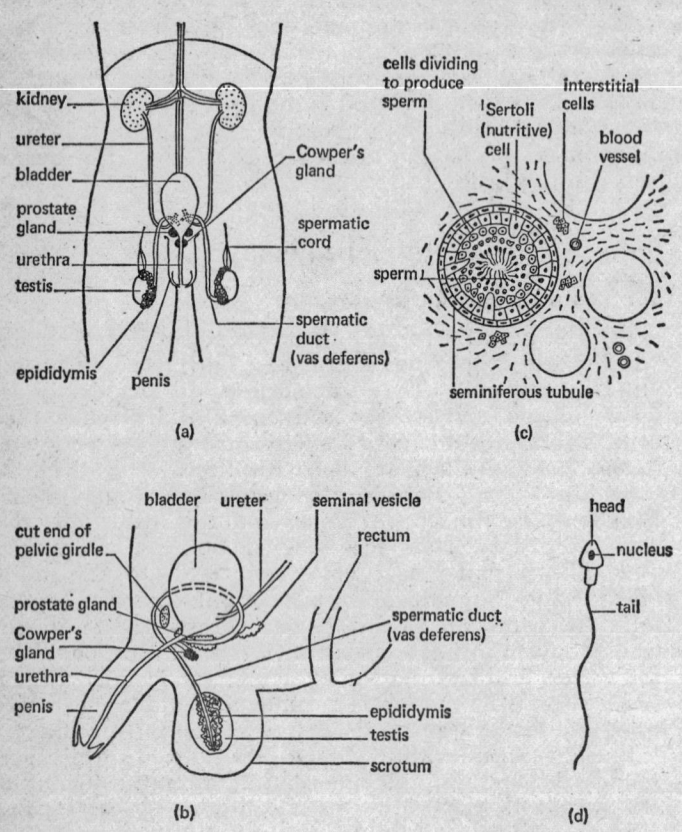

FIG. 58. The male reproductive system: (a) male reproductive system (front); (b) male reproductive system (side); (c) section through testis; (d) sperm.

stiff so that it is possible to insert it into the vagina of the female (*see* Fig. 58).

11. The female reproductive system.

(*a*) The paired *ovaries* lie attached to the abdominal wall below the kidneys. They consist largely of connective tissue through which blood vessels run. Near the surface are some 70,000 potential egg cells. There is no continuous production of eggs during the life of the individual. At puberty (11–16 years old) the egg cells mature and are discharged from the ovaries. They are produced from alternate ovaries at 28-day intervals. This process continues until a woman is 45–50 years old, so that only some 500 egg cells ever reach maturity.

(*b*) *Development of the egg.* The cells around the egg cell divide and become infiltrated by blood vessels. These cells enlarge so as to enclose the egg in a fluid-filled sac called the *Graafian follicle.* When the follicle is fully developed it is about 15 mm in diameter, and the egg is about 0·2 mm in diameter. At this stage it migrates to the surface of the ovary and bursts, discharging the egg into the oviducal funnel. The funnel is lined with cilia which wafts the egg down the oviduct. After 3–4 days it reaches the uterus. After the egg is discharged the cells in the Graafian follicle develop to form the yellowish *corpus luteum* which is a temporary gland.

(*c*) *The other organs of the female reproductive system.* Around each ovary, but not joined to it, is the *ciliated funnel* of the *oviduct.* Each oviduct is a coiled tube which runs laterally to join the *uterus.* The uterus (womb) has a thick wall, and it is in it that the embryo develops. At the lower end of the uterus is a ring of muscle—the *cervix*—which closes off the uterus from the *vagina* with which it joins. The vagina is a muscular tube which opens to the exterior through the *vulva.* In the female, the urethra is not part of the reproductive tract, but opens independently into the vulva (*see* Fig. 59).

12. The menstrual cycle.

(*a*) *Menstruation.* The ultimate fate of the fertilised egg is to become embedded in the wall of the uterus where it

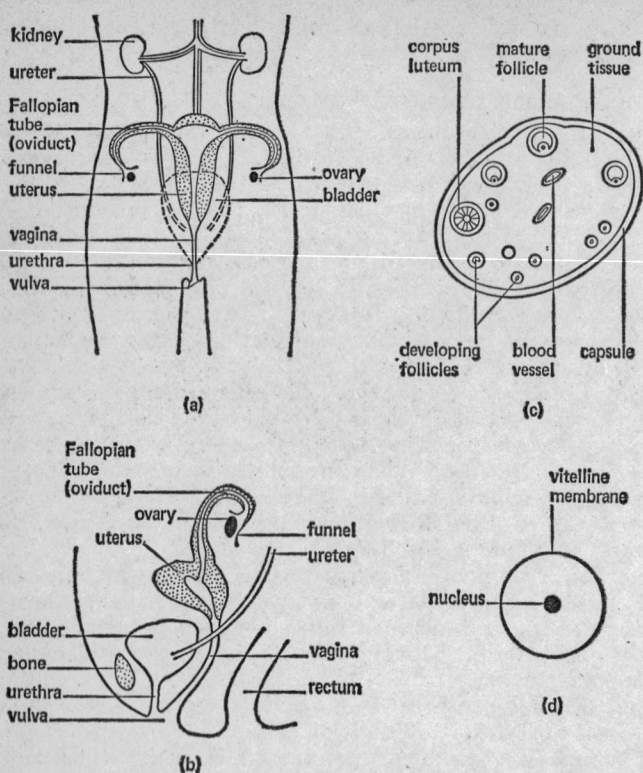

FIG. 59. The female reproductive system: (a) female reproductive system (front); (b) female reproductive system (side); (c) ovary; (d) human egg.

will begin its development into an embryo. So that the uterus will become prepared to receive the egg it becomes greatly thickened and perfused with blood. This thickening is co-ordinated by hormones, with the discharge of the egg. If the egg is not fertilised, it, and the thickened tissue of the uterus wall, with blood and mucus, are discharged through the cervix and vagina. This process is called *menstruation*.

It occurs approximately every 28 days and is referred to as the monthly period. The whole cycle of events which takes place between the beginning of one menstrual flow and the next is the *menstrual cycle*, ovulation taking place on about the 14th day.

(*b*) *Co-ordination of the menstrual cycle*. The menstrual

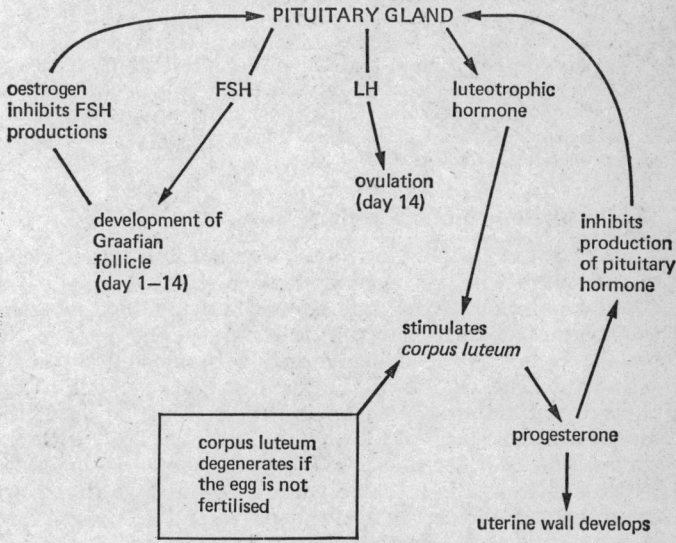

FIG. 60. Co-ordination of the menstrual cycle.

cycle is controlled by three hormones produced by the anterior lobe of the pituitary gland. These, in turn, stimulate various parts of the ovary to produce hormones:

(*i*) *Follicle stimulating hormone* (*FSH*) causes the Graafian follicle to mature and produce the hormone *oestrogen*.

Oestrogen: a. Inhibits the production of FSH ensuring that only one follicle develops at a time; and

b. is responsible for the development of the secondary sexual characteristics during adolescence and their maintenance during later life.

(*ii*) *Luteinising hormone* (*LH*) stimulates the Graafian follicle to extrude the egg during the 13–14th day of the cycle.

(*iii*) *Luteotrophic hormone* stimulates the corpus luteum to produce *progesterone*. Progesterone has two effects:

a. Prevents the production of further sex hormones by the pituitary gland (hence indirectly preventing the production of any more eggs); and

b. stimulates the uterus lining to thicken to receive the fertilised egg. If the egg is not fertilised, the corpus luteum degenerates, so that the uterine wall is restored to normal, menstruation occurs, and the cycle is repeated. If an egg is fertilised, the corpus luteum persists and continues to produce progesterone. This blocks the production of the pituitary hormones, so that the cycle stops, resulting in pregnancy (*see* Fig. 60).

13. Copulation and fertilisation.

(*a*) *Copulation*. Fertilisation is internal, the actual fusion of the sperm with the egg takes place in the oviduct. The act of copulation (sexual intercourse) is the method whereby the spermatozoa are placed in the uterus. The penis of the male becomes erect, being gorged with blood, and is inserted into the vagina of the female. Sensory cells in the penis are stimulated, causing the reflex contraction of the muscles of the sperm duct and epididymis. This, with the contraction of other muscles, forces the semen through the urethra of the male into the vagina. From here the sperm swim through the cervix into the uterus and up the oviduct. If copulation has occurred when there is an egg in the oviduct, fertilisation may result.

(*b*) *Fertilisation*. During copulation the male discharges 200–300 million sperm, but only one of these fuses with an egg. The egg produces a substance which attracts the sperm to it. One of the sperm becomes attached to the surface of the egg. The tail is left outside and the head passes through the protoplasm of the egg to fuse with the nucleus. Simultaneously with the passage of the sperm head into the egg, a membrane develops around the egg which prevents the entry of other sperm into it.

GROWTH AND DEVELOPMENT OF THE EMBRYO

14. Early cell divisions, and implantation.

(*a*) *Early cell division.* Immediately after fertilisation, the zygote begins to divide (*cleavage*) first into 2, then 4, 8, 16, 32, etc. to form a hollow ball of cells (*blastocyst*). These divisions take place as the blastocyst is passing down the oviduct. The directional movement in the oviduct is brought about by its muscular contraction and a flow of fluid through it.

(*b*) *Implantation.* Before implantation the blastocyst is provided with food and oxygen from the fluid in which it is floating, but later it becomes embedded in the uterine wall. As the embryo becomes embedded in the uterine wall (*implantation*) it develops finger-like villi which penetrate through the outer wall of the uterus (*endometrium*) into blood sinuses within the wall. During this stage the wall becomes thicker and more infused with blood.

15. The placenta and later development.
The villi from the embryo become greatly extended over the wall of the uterus to form the *placenta*. At a later stage the umbilical cord runs from the placenta to the developing embryo.

(*a*) *The placenta* is a disc of tissue when it is finally developed. It has numerous villi passing from it into the blood sinuses of the uterus.

NOTE: At no stage does the maternal blood mix with that of the embryo. The membranes separating them function selectively, preventing harmful materials from passing to the embryo from the mother. The separation of the two blood systems is also necessary because the higher pressure of the maternal blood would damage, or even break, the fine capillaries in the embryonic system.

The membranes separating the two blood systems are sufficiently permeable to allow dissolved glucose, amino acids, salts, and oxygen to pass from the maternal blood to that of the embryo, and to allow carbon dioxide and waste nitrogenous substances to diffuse in the opposite direction (*see* Figs. 61, 62).

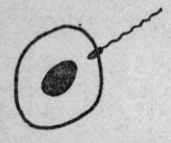

(a) Sperm penetrates egg

(b) Head of sperm (with nucleus) moves to egg nucleus

(c) Male nucleus penetrates female nucleus

(d) 2 - celled stage

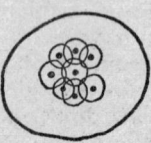

(e) 8 - celled stage

(f) Blastocyst

FIG. 61. Fertilisation and early development.

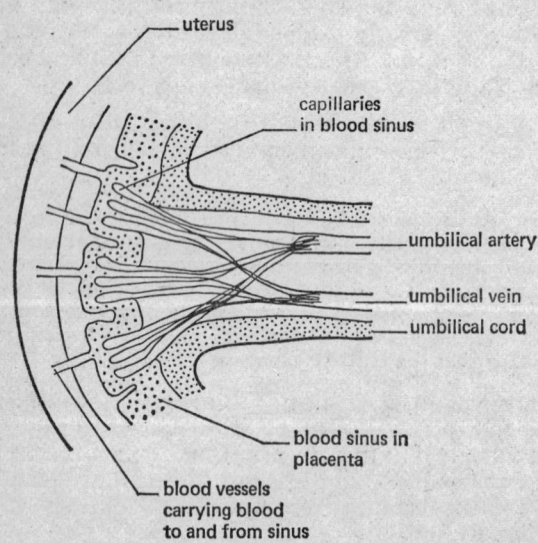

FIG. 62. Blood circulation in the placenta.

(b) Later stages in development.

    (i) Fertilisation.

    (ii) Ten days—implantation complete.

    (iii) Fourteen days—embryo forms a small lump in the uterine wall, about 12 mm in diameter.

    (iv) One month—spinal cord developed, and the head is recognisable.

    (v) One to two months—development of the amnion complete. The embryo 10 mm long, and the limbs recognis-

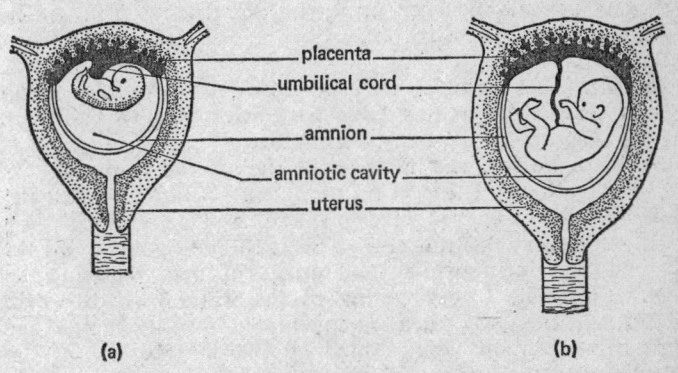

placenta

umbilical cord

amnion

amniotic cavity

uterus

(a)   (b)

FIG. 63. Development of the embryo: *(a)* 4–5 weeks; *(b)* 5 months.

able as small lumps. The skeleton developing rapidly. The blood system is developing and the heart has two chambers.

    (vi) Two to three months—the placenta has formed during the first part of this period. By the end of it all the organs are formed, but are not fully developed. The embryo weighs about 14 g.

    (vii) Four months—at the end of the period the embryo weighs about 110 g.

    (viii) Five months—the baby begins to move its limbs.

    (ix) Six months—by the end of the time the air sacs have developed and the nostrils are opened. The baby weighs about 1 kg.

    (x) Seven months—at the end of the 7th month, development is complete. The baby weighs about 2 kg, the increase in weight being due partly to the depositing of fat under the

skin. He also turns over so that the head lies immediately above the cervix, ready for birth.

(*xi*) *Nine months*—by the end of the 9th month, the finger nails and toe nails are fully grown. The baby is about 50 cm long and weighs 3½ kg. The baby is ready to be born.

(*c*) *Embryonic membranes*. During the early stages of development, outgrowth from the embryo grow around it to form a protective sac. The most important of these membranes is the *amnion*. The cavity of the amnion is filled with fluid in which the embryo is able to move freely. The fluid acts as a shock-absorber, protecting the embryo from mechanical damage (*see* Fig. 63).

16. Birth. As the baby has been growing, the muscles of the uterus, cervix and vagina have been stretching to allow for the free passage of the baby during birth.

As birth begins, the muscles of the uterus begin rhythmic contractions. These begin at about 15-minute intervals, gradually increasing in strength and frequency until they occur once every 3–5 minutes. This is the first stage of *labour* at some time during which the amnion ruptures, releasing the fluid it contains (the breaking of the waters). The cervix dilates, and the baby's head is now forced into the vagina, the uterine contractions being aided by the contractions of the muscles of the abdomen. Finally, the baby passes through the vagina to the outside world.

Some time after the birth of the baby, the placenta and the remains of the membranes pass out through the vagina as the "after birth".

The time taken for the birth process varies considerably, but is usually about 3 hours.

The first action of the baby on being born, is to cry. This is a reflex initiated by the drop in temperature it experiences. It is essential so that air is drawn into the lungs.

When the baby is born, it is still attached to the placenta by the umbilical cord, which has to be cut after birth. The pulse in the umbilical cord blood vessels soon stops, and the cord is severed. It is first ligatured about 5 cm from the baby by two ligatures, and the cut made between them. After a few days, the remains of the cord attached to the baby shrivels and falls off to leave a scar—the *navel*.

NEEDS OF THE PREGNANT WOMAN

17. Diet. During her pregnancy, the mother has to produce about 3½ kg of new cell material in the form of her baby. To do this she requires a diet rich in protein and minerals, especially calcium for the formation of bones. A diet containing meat, eggs and milk is important. Milk is essential as it is an excellent source of calcium. Lack of calcium can cause degeneration of the mother's bones and teeth as the reserves are withdrawn to feed the baby. An increase in carbohydrate intake is not desirable as it will cause an increase in weight.

A balanced supply of vitamins is also necessary, so that the diet should include plenty of fresh vegetables, yeast extract and wholemeal bread.

The importance of the diet to expectant mothers is recognised by many governments, so they provide vitamin tablets, orange juice, etc. at reduced cost.

18. Clothing. Clothing should be loose and comfortable, so as not to restrict the blood supply. Low-heeled shoes should be worn, as high-heeled shoes throw the body out of balance. This is accentuated as pregnancy advances, and may lead to back-ache.

19. Rest and exercise.

(a) *Regular mild exercise, e.g.* walking, is valuable as it keeps all the muscles in tone, and stimulates the circulation. Violent exercise, lifting heavy weights and stretching should be avoided as they may strain the already distended abdominal muscles.

(b) *The pregnant woman needs plenty of rest.* An afternoon rest is advisable, preferably lying down with the feet slightly raised. This helps to prevent varicose veins from developing. The afternoon rest should be continued after the baby is born, especially when it is being fed during the night.

20. Personal hygiene.

(a) The blood system of an expectant mother is removing waste products from the baby as well as her own, so that she should drink plenty of water to ensure that toxins from both sources are removed from her system.

(b) Toxins can also accumulate as a result of constipation and from decaying teeth. Regular evacuation of the bowels and regular dental treatment throughout pregnancy are important. In some countries, e.g. the U.K., the state provides a free dental service for expectant mothers.

(c) Alcohol is a poison and should not be taken during pregnancy.

(d) There is strong evidence to support the view that smoking is detrimental to the health of the unborn child. Babies of women who smoke tend to be smaller than those of women who are non-smokers.

21. Attitude of mind. Pregnancy is a great emotional experience as well as a physiological one, and the mother-to-be is frequently frightened by it. This attitude is often exaggerated by "old wives' tales" and stories told by other women. It is essential that the pregnant woman realises that having a baby is a perfectly natural process and that her body is anatomically and physiologically adapted to it. Regular attendance at pre-natal clinics will help a lot to prepare a woman to have her baby. Sound advice will be given, and the mother will be taught exercises which will help to make the birth of her baby easier.

CARE OF THE BABY

22. Feeding.

(a) *For the first 4 months* the baby is fed on milk. It may be breast-fed or bottle-fed. Most mothers breast-feed their babies, at least at the beginning, but sometimes, as the baby's demands increase, the mother's milk supply may not be sufficient. The advantages of breast-feeding are:

(i) The baby is povided with antibodies and mineral ions in the correct balance by the mother;
(ii) the mother and baby benefit psychologically from their intimate relationship.

Neither the mother nor the baby suffers if the baby is bottle-fed correctly, but breast-feeding is to be encouraged.

Vitamin C (as rose hip syrup) and vitamins A and D (as cod liver oil) can be given after about three weeks.

The baby is fed at 4-hourly intervals at the beginning, but as the food intake increases, it gradually manages to do without the 2 a.m feed.

(b) *At about* 4 *months old*, a little (about half a teaspoonful) cereal is added to the diet. This is gradually increased, until about 6–7 months old, the baby is having 4 teaspoons of cereal twice a day. At about 6 months, strained stewed fruit, eggs, strained meat and vegetables are added to the diet, beginning with a teaspoonful, and increasing to about 6. All these additional foods are not given at each meal, *e.g.* strained meat followed by milk pudding may be given at mid-day; the yolk of a lightly boiled egg may replace the meat, or fruit replace the milk pudding.

(c) *By* 6–8 *months old* the baby will not need the 10 p.m. feed. At this stage a three-meal-a-day routine can be established. The amount of solid food is increased, and the amount of milk decreased. The baby is becoming active, and from now on the food intake will increase. It is important to avoid feeding a child an excess of carbohydrate, *e.g.* bread, potatoes, sweets. The child puts on too much weight and the appetite for body-building proteins is lost.

23. Sleep.

(a) A young baby will sleep throughout the 24 hours when it is not being fed.

(b) Towards the end of the first year, it should sleep 14 hours a night, an hour in the morning and an hour during the afternoon. This will continue to 3–4 years old when the day-time rest periods are gradually eliminated.

(c) Children up to 10 years old need about 12 hours sleep a night. This is gradually reduced until adulthood when about 8 hours a night is needed.

24. Clothing.

(a) A child's clothes should be loose fitting to allow free movement and circulation, but not so loose as to cause rubbing and chafing of its tender skin. If the clothes are too tight the baby is uncomfortable and irritable. If the circulation is restricted, it can make the baby cold.

(b) Babies' outer clothes, especially those of toddlers, get dirty quickly. Therefore they should be made of an easily washable material, e.g. nylon or other man-made fibre. Clothes next to the skin should be sweat absorbent and porous for comfort and warmth. Natural fibres are probably best.

(c) Bedding should be light and warm. A sleeping bag is ideal.

GROWTH AND DEVELOPMENT OF MATURITY

25. Introduction. Figure 64 shows the rate at which a human being grows. The most rapid increase in height is during the first 2 years. Between the second year and the onset of

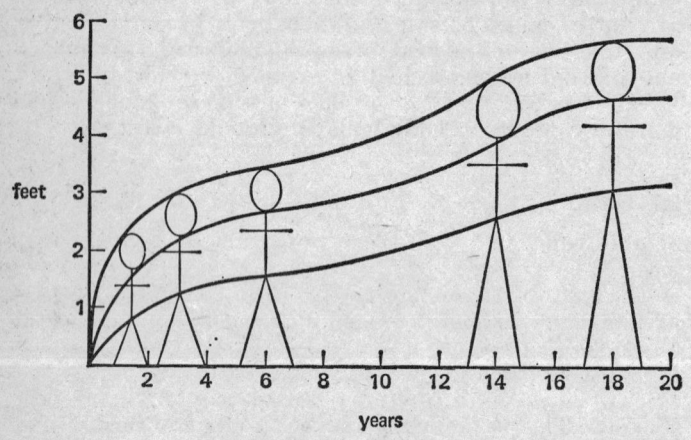

FIG. 64. Human growth curve.

puberty the growth rate is uniform. At the onset of puberty there is often a spurt of growth. Notice the relative proportions of the parts of the body during the different stages of development. The head of a new-born baby is about one-quarter the total length of the body, while the legs are about the same length. The head grows less in comparison to the rest of the body, so that in the adult it is a little over one-eighth the total length, while the legs are about one-half the length.

26. The early years. A new-born baby weighs about 3 kg and is 50 cm long. By the time it is 6 months old it has doubled its weight, and by the end of the first year the birth weight is trebled and the length increased to about 70 cm. By the end of the first year it will have begun to walk. (The time taken to begin to walk is very variable.) After the second year the legs begin to lengthen and the growth rate of the head and trunk slows. Between 5 and 7 years old is a period of rapid growth and muscular development.

27. Puberty. At this stage the sex organs mature. In girls it takes place at 11–16 years old. Menstruation begins, hairs develop under the arm-pits and in the pubic region; the breasts develop, and deposits of fat under the skin make the girl's figure more like that of a young woman.

At 10–14 years old pubic development begins in boys. Hair develops under the arms and in the pubic region, and on the face and chest. The shoulders broaden and the larynx enlarges so that the voice deepens (breaks). This period is one of emotional as well as physical development, and the individual moves from childhood to become an adult.

28. Ageing. From the age of about 40 years, the body metabolism slows down and the individual becomes less physically and mentally active. Cell activity slows and their rate of division decreases. Because of this, wounds heal less readily, and the body becomes more readily infected. The senses become less acute. In women, the ovaries gradually stop producing eggs after 40–50 years old. Menstruation ceases. This stage in development is called the *menopause*.

In the U.K. women retire at 60 years old, and men at 65, but in these days of modern medicines and sensible diets, people of this age are still quite active and go on to enjoy long periods of retirement.

PROGRESS TEST 8

1. Define sexual reproduction.   (2)
2. What are the disadvantages of sexual reproduction?   (2)
3. What happens to the number of chromosomes during meiosis?   (3)
4. Why is meiosis important in the formation of gametes?   (3)
5. What is the importance of crossing over?   (4)
6. What is the function of DNA in the nucleus?   (5)
7. What is the function of RNA?   (5)
8. What is the monohybrid ratio?   (6)
9. Explain what is meant by the term "dominant".   (6)
10. Define the terms "genotype" and "phenotype".   (6)
11. Define the terms "homozygous" and "heterozygous".   (6)
12. What is meant by "incomplete dominance"?   (6)
13. Explain the term "dihybrid ratio".   (6)
14. Draw a diagram to show the functioning of the sex chromosomes in determining the sex of an individual.   (7)
15. Define the term "mutation".   (9)
16. Draw and label a diagram of the male reproductive system.   (10)
17. Draw a diagram of the female reproductive system.   (11)
18. State two functions of FSH.   (12)
19. State two functions of progesterone.   (12)
20. What is the difference between copulation and fertilisation?   (13)
21. What are the functions of the placenta?   (15)
22. What are the functions of the embryonic membranes?   (15)
23. Name three substances essential in the diet of a pregnant woman.   (17)
24. Why is mild exercise valuable to a pregnant woman?   (19)
25. State four points of personal hygiene which should be observed by a pregnant woman.   (20)
26. What are the advantages of breast-feeding.   (22)
27. How long does it take for a baby to double its birth weight?   (25)

EXAMINATION QUESTIONS

1. State Mendel's First Law. Explain why two brown-eyed parents can have a blue-eyed baby, *and* why there is an equal chance of the baby being a boy or a girl.
2. Make a labelled drawing of the female reproductive system. What are the functions of: (*a*) the placenta; and (*b*) the embryonic membranes?

3. What part is played by the female sex hormones during: (a) puberty; and (b) the oestrus cycle?

4. Write notes on the following: (a) genes; (b) DNA; (c) RNA; (d) crossing over.

5. Discuss the needs of a pregnant woman under the following headings: (a) diet; (b) clothing; (c) special medical treatment in clinics.

CHAPTER IX

THE RELATION OF MAN TO OTHER ANIMALS

MULTIPLICITY OF LIFE

1. Mammals. Man is an animal and a member of the group of animals called the *mammals*.

Mammals all have four limbs (quadrupeds) which are penta-dactyl. They breathe by lungs. The heart is four-chambered. They are *homoiothermic* (warm-blooded). The body is covered by hair and the young are suckled by mammary glands. Be-cause man is a mammal he is related to animals such as the horse, lion, dog, rat and the various monkeys.

2. Other types of animals. Mammals are highly complex animals, but others, *e.g. Amoeba*, are non-cellular, while yet others are multicellular. The simple multicellular animals include the *Coelenterata, e.g.* sea anemones, *Hydra*. More complex multicellular animals include such animals as tape-worms and liverflukes (*Platyhelminthes*), earthworms (*Anne-lida*), crabs, etc. (*Crustaceae*), insects (*Insecta*) and spiders (*Arachnida*). This list does not include all the types of animals, but will give you some idea of the variety of the group of animals called the *invertebrates* because they have no back-bone.

The other large group of animals, the *vertebrates*, have a backbone. The vertebrates include the fishes, amphibians (*e.g.* frogs and toads), reptiles (snakes, crocodiles, lizards), birds, and mammals. Only the birds and mammals are homo-iothermic.

3. Types of plants. The other great group of living organ-isms is the plants. This includes the *Bacteria*, Seaweeds and the green slimes of ponds, etc. (*Algae*), moulds and mushrooms (*Fungi*), liverworts and mosses (*Bryophyta*), ferns and their allies (*Pteridophyta*), cone-bearing trees, etc. (*Gymnosperms*), and the flowering plants (*Angiosperms*). The green plants photosynthesise, manufacturing sugars from carbon dioxide and water using light as a source of energy. The green pigment *chlorophyll* is essential for this process. All animals are ultimately dependent on green plants for their food, so the green plant is a very important life form.

EVOLUTION

4. Historical background. Ever since the times of the Ancient Greeks, man has been looking for some relationship between the vast number of different plants and animals. He was aware of the similarities between groups, and their differences from others, and used these to classify the various organisms. The greatest system of classification was advanced by the Swedish biologist Linnaeus in the eighteenth century, and his system is the same one as is used today. Although systems of classification developed, the idea that one form of organism had actually changed to produce a new form, *i.e. evolved*, made little headway. This lack of progress was due in part to the teaching of the Church. The Bible was treated as being absolutely true, therefore the story of the Creation in the Book of Genesis was taken as a fact. In this it is stated that God created all the forms of life separately, and man in His own image (the theory of Special Creation). Therefore the idea that animals and plants had evolved one from another was ludicrous, and the concept of man evolving from other mammals was heresy.

In 1859, Charles Darwin published a book called *On the Origin of Species by Natural Selection*, and in the same year Alfred Russell Wallace published a paper containing the same ideas. These two scientists put the theory of evolution on a scientific basis, and after a tremendous controversy with the religious leaders of the time, it was gradually accepted as being the most plausible theory of how plants and animals now occur in such diverse forms.

The theory can be stated briefly as follows:

(*a*) Plants and animals reproduce at a very high rate, so that if all the offspring survived, they would soon exhaust the available food supply.

(*b*) Because of this there is a "struggle for existence"

from which only the best-adapted organisms survive to breed. These variations may be very slight in any particular generation, but as the process is repeated through successive generations, the characteristics of the offspring become modified to such an extent as to justify them being grouped as a species different from the original parents.

5. Evidence.

(a) *Fossils.* Fossils are the impressions of organisms left in rocks. They may be just the shape, left by parts which have since decayed, or the skeletons which have become impregnated with minerals. When an aquatic animal died, it fell to the bottom of the sea. Over the years it would be covered by sediment which became compressed to form rock. During geological changes, these rocks have become exposed to reveal the fossils. Plants and animals that did not live entirely in water would have become covered with water and mud when they died. The oldest rocks contained the oldest fossils, so that by examination of these fossils we can obtain a picture of the living organisms throughout the ages. The fossils show a definite relationship with each other throughout the scale, lending support to the idea that one form has evolved from another.

Table V opposite shows the relationship between the ages of some of the rocks and the life forms found in them. Notice the time scale, and see how little time man has been on the earth compared with the other animals.

(b) *Variety within groups of organisms, e.g.* vertebrates all have backbones and, with the exception of the fishes, they all have pentadactyl limbs. Vertebrate teeth have similar structures. Mammals have all the characteristics mentioned earlier, but when we consider the species, *e.g.* dogs, wolves and jackals are all in the same species, they have characteristics in common which would distinguish them from other mammals.

(c) *Embryology.* Vertebrate embryos, when compared during the early stages of development, are remarkably similar. For example, the embryos of a fish, bird, and man all have developing gills. (They only develop to the early stages in the bird and man.)

(*d*) *Biochemistry.* Similar chemicals occur in different genera of animals, *e.g.* the thyroid hormone from the sheep can be used to treat thyroid deficiency in man; keratin occurs in feathers and hairs.

TABLE V. GEOLOGICAL PERIODS AND LIFE ON EARTH.

Time in million years	Era	Period	First appearance
1	Pleistocene Holocene	Quaternary	Man
10 30 40 60 75	Pliocene Miocene Oliocene Eocene Paleocene	Tertiary	Flowering plants
140 170 200	Cretaceous Jurassic Triassic	Mesozoic	Mammals Birds and early Mammals
220 275 320 350 420 520	Permian Carboniferous Devonian Silurian Ordovician Cambrian	Palaeozoic	Conifers Reptiles Amphibians Early fern relatives Insects Fish Many invertebrates
—	Precambrian		Algae

6. Evolution of man. We must be careful not to fall into the trap of saying that "man has evolved from apes". It is more likely that man and apes have evolved from a common ancestor. This view is supported by the evidence of fossils.

The earliest known ape was probably an animal called

Proconsul. It lived about 30 million years ago. Various other ape-like fossils have been discovered which suggest the link between it and the apes. The oldest man-like animal to be discovered was *Homo habilis.* He lived about 1 million years ago. He was about 1 m tall and made crude tools out of stone. His brain was much smaller than that of modern man. The next link in the evolutionary chain that has so far been discovered is *Homo erectus.* He lived about half a million years ago and was skilled in making and using tools. *Neanderthal man* lived between 70,000 and 40,000 years ago while *Pekin* and *Java man* lived about 350,000 years ago. These three types of man made tools and lived in caves. They also used fire. Their brains were about the same size as modern man's (Neanderthal man's brain was larger), but for some reason all became extinct. *Cro-Magnon man* lived about 40,000 years ago and is the first representative of modern man, *Homo sapiens.* He was tall, even by modern standards, walked erect, and had a large brain; he looked very like modern man, but through some unknown cause, became extinct.

7. The distinctive features of man.

(*a*) *Large brain size.* The brain size (*cranial capacity*) of man is about 1,600 cm³. That of an ape is 600 cm³.

(*b*) *Ability to oppose the finger and thumb.* This makes it possible to hold and manipulate objects.

(*c*) *Extensive manufacture of complex tools.*

(*d*) *Development of a complex language,* which enables him to communicate ideas.

(*e*) The increased efficiency in catching prey (and ultimately domesticating animals) gave time for *leisure.* This allowed for the development of various forms of art and culture, beginning with cave paintings.

(*f*) Of all the animals, man is the only one that has any *idea of religion.* Even the most primitive tribes worship a god and have some idea of life after death.

PROGRESS TEST 9

1. Give five features that distinguish mammals from other animals. **(1, 2)**

2. Why is chlorophyll essential to all life? **(3)**

3. What are the two main points of Darwin's theory of evolution? (4)

4. On what four main groups of evidence did Darwin base his theory? (5)

5. How long ago did the first man-like creature live, and what is he called? (6)

6. How long ago did the first modern man (*Homo sapiens*) live? (6)

7. Give six distinctive features of modern man. (7)

EXAMINATION QUESTIONS

1. Write an essay to explain Darwin's theory of evolution.
2. Write an account of the evolution of man.

CHAPTER X

MAN AND OTHER ORGANISMS

DEPENDENCE OF MAN ON PLANTS

1. Man is the great consumer.

(*a*) *All animals depend on plants for food*. The inter-relationship has been mentioned in II where food chains are discussed. The staple diet of man is various forms of plants. In the West wheat is the chief food. It is used to make flour for bread, macaroni, etc. In the East rice is the staple cereal. As it is unsuitable for making flour, it is usually eaten boiled in various ways. In areas where the climate is suitable, maize (corn) is the chief grain crop, while in some of the drier areas sorghum and millet are used.

The underground parts of plants feature largely in the diet. There are a great number of these that are of local importance. Potatoes are used largely in the West, but in tropical countries yam and taro are of local significance. These are all sources of starch. Sugar beet is an important source of sugar.

Apart from these basic foods, large acreages of the world are devoted to growing fruits, *e.g.* citrus fruits, grapes, apples; beverages, *e.g.* tea, coffee and cacao; and spices, *e.g.* cloves, for man to eat.

Huge acreages of the earth's surface are devoted to the growing of timber for making paper and for building and furnishing houses. The rarer timbers are felled in the tropical forests where they grow in their wild state.

Much of the earth's surface is covered with grassland, which is utilised to feed livestock for man's use. The prairies of America and the sheep lands of Australia and New Zealand are examples.

Man grows several other plants for his use, *e.g.* cotton, flax, rubber.

Man is also dependent on the plants growing in the sea for

his food. He eats fish, which are the end consumers in food chains which begin with the small algae floating in the sea.

(*b*) *Man alters the natural vegetation to provide himself with plant products.* To make the land suitable for the production of crops, the natural vegetation is destroyed, *e.g.* grassland ploughed for wheat, or forest destroyed for rubber plantations. This may destroy the delicate balance which has been established over the years between the vegetation, climate and soil. If this balance is destroyed, the soil is eroded or its structure altered so that it becomes a useless wilderness. It is the function of good agriculture to preserve the soil as well as produce crops from it.

(*c*) *Man is using up the plant reserves formed in ages past.* Coal and oil are dead plant material. They are the partially decomposed remains of plants that lived millions of years ago, *e.g.* the plants from which coal is formed were growing during the Carboniferous period. Coal, oil, and natural gas are now being exploited so rapidly, that within the foreseeable future the supplies will be exhausted and we will have to use alternative sources of power, *e.g.* nuclear energy, solar (sun) energy, wind and water.

(*d*) The production of food requires the *utilisation of other minerals*. To meet the increasing demand for food, fertilisers have to be used to replace the minerals taken from the soil by plants. Most of these fertilisers are made from chemicals, which in turn are made from minerals extracted from the ground. For example, most of the sulphur produced is used to make sulphuric acid, which in turn is used to make the fertilisers superphosphate and ammonium sulphate.

BACTERIA

2. Distribution. Bacteria occur in every part of the earth where it is possible for living organisms to exist. Many are saprophytic, living in soil and water and in any moist organic matter. Other saprophytes live in human food, *e.g.* milk and meat, which they decompose.

Others are parasites on plants and animals, their presence resulting in disease. Yet others can photosynthesise in a way similar to that employed by green plants. Some species can use inorganic chemicals as a source of energy.

3. Structure and reproduction.

(a) It is not strictly accurate to speak of a bacterial cell, because it is structurally different from that of a plant or animal. The cell wall is more chemically complex than that of a green plant. Although the cells contain DNA and RNA, these are not organised into a nucleus surrounded by a membrane.

(b) Bacteria are identified by their shapes, and their chemical activity. Fig. 65 shows some of the shapes. Some have flagella which they use to swim. The *bacillus* is rod-shaped; *cocci* are spherical and may be single or in groups; the *spirillum* is twisted; and the *vibrio* is comma-shaped and slightly twisted.

(c) *Bacterial reproduction.*

(i) The most important method of reproduction is by the simple *splitting of the cell*. A cell grows normally to its maximum size and splits into two.

(ii) Sexual reproduction has been observed in some bacteria. Two cells fuse together and then divide asexually. This method of reproduction is of little importance in building-up a population, but may be important genetically.

(iii) Some bacteria are capable of forming a resistant coat (*endospore* or *cyst*) which is capable of withstanding extremes of temperature and desiccation. When the spore is again in favourable conditions, it will germinate to produce a normal cell.

Bacteria are capable of very rapid reproduction. Under favourable conditions a cell will divide three times in an hour. Such rapid reproduction is balanced by a rapid death-rate. A rapid increase in population leads to an accumulation of poisonous biproducts (toxins) which kill the bacteria. Parasitic bacteria may kill their host. All those that are not spore-formers are very susceptible to drought, so drying out of the medium on which they live can kill them. Many bacteria are killed by ultra-violet light. This is present in sunlight.

In considering our own health and the preservation of food it should be remembered that dark, warm, dirty, damp con-

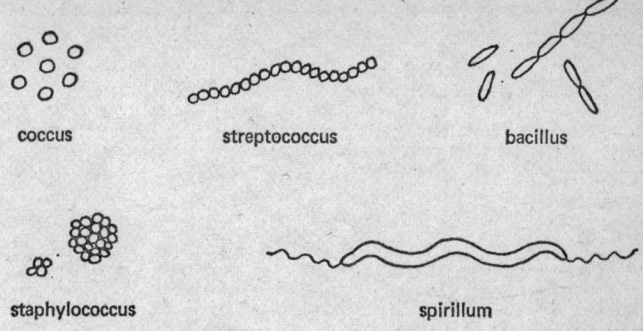

FIG. 65. Types of bacteria.

ditions are ideal places for the multiplication of bacteria (*see* Fig. 65).

4. Beneficial bacteria.

(*a*) *Soil bacteria*. The soil bacteria are responsible for decomposing the dead remains of plants and animals and their excreta. Other organisms, *e.g.* fungi, also play a part in this decomposition. As a result of this decay, which is similar to the digestive processes of human beings, the bacteria release simple compounds from the complex ones and make them available for absorption by plants. Thus, the supply of materials for the nutrition of plants and animals is re-cycled. Two specific cases of this are the *carbon cycle* and the *nitrogen cycle*.

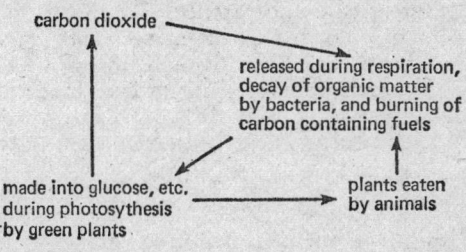

The carbon cycle

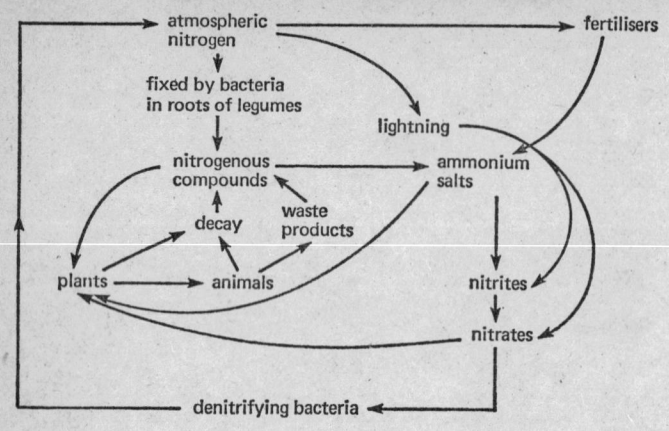

The nitrogen cycle

All the stages in the nitrogen cycle, except the addition of fertilisers and the fixation of atmospheric nitrogen by lightning involve different types of bacteria.

During the rotting of organic matter in the soil, *humus* is formed. This improves the structure of the soil, making it more suitable for plant growth.

(b) *Sewage disposal.* One of the chief problems that faces mankind is the disposal of his own excreta. Faeces are a dangerous source of disease-causing bacteria. In areas of high-density population it is imperative that this waste material, together with domestic and industrial waste, is disposed of effectively and rapidly. To do this man utilises bacteria to break down this offensive waste into harmless substances like carbon dioxide and water which can be returned to the environment. This process is carried out in a *sewage works*.

The two basic processes carried out in a sewage works are:

    (i) *Filtration;*
    (ii) *decomposition* by micro-organisms.

There are two main methods used in sewage works. In both methods the primary treatment is the same, namely:

(*i*) Screening through the metal grid to trap large pieces of material, *e.g.* paper and rags. The material is removed from the grid mechanically and the debris is burnt or buried.

(*ii*) Slow passage through a narrow open channel to allow grit and stones to settle out.

(*iii*) After this the sewage may:

a. *Either* pass into sedimentation tanks in which the solid matter is allowed to settle as *sludge*. Sedimentation may be accelerated by adding aluminium sulphate or iron (III) chloride. The sludge is then removed and digested in aerated tanks. In this process, bacteria break down the solid organic matter to methane and water. The methane can be used as a fuel. Any residue is dried and used as a fertiliser.

b. *Or* the raw sewage is mixed with a little sludge (containing bacteria) and aerated by bubbling air through it. The organic matter is oxidised. Any residue which settles is dried and used as a fertiliser. This is called the *activated sludge method*.

(*iv*) The water left by either process still contains organic particles, and is filtered through beds of clinker. The water is sprinkled on to the clinker through rotating sprinklers. Bacteria on the surface of the clinker digest any organic matter. The water, which is now free of organic matter, is released into a river (*see* Fig. 66).

(*c*) *Vitamin formation.* The bacterium *Acetobacter suboxydans* is used in the commercial preparation of vitamin C (ascorbic acid from the sugar sorbitol). Vitamin B_{12} is produced in the intestine of man by bacteria.

(*d*) *Lactic acid production.* The production of lactic acid by *Streptococcus lactis* makes milk sufficiently acid for it to form the curds necessary for *cheese-making*. Lactic acid is produced commercially for the food industry by the fermentation of sugars, using species of *Lactobacillus*.

(*e*) *Silage.* Various bacteria and other organisms play a part in the fermentation which produces the acids which preserve silage, and make it palatable to livestock. (Silage is fermented green fodder—a method of preserving greenstuff for winter feeding to stock.)

(*f*) *Retting flax.* Various bacteria produce the enzyme pectinase which breaks down the tissues of the flax stem (retting) when they are wet. This separates the fibres which are used to make linen.

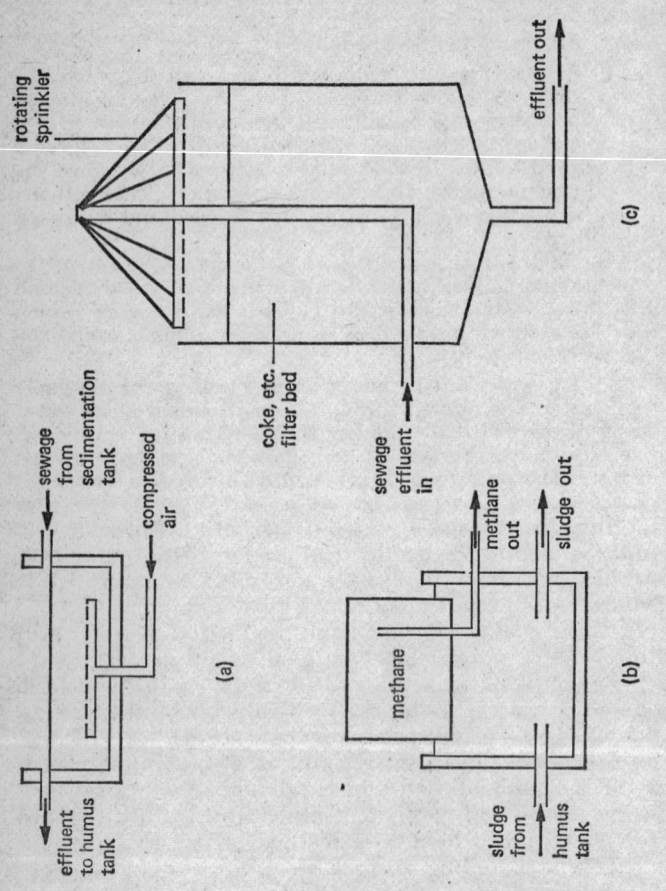

Fig. 66. Sewage disposal: (a) aeration tank; (b) sludge digestion tank; (c) percolating filter.

(g) *Biological washing powders.* These contain proteolytic (protein-destroying) enzymes which are extracted from bacterial cultures.

5. Pathogenic bacteria.

(a) *Spread of pathogenic bacteria.* As bacteria are very small and may form spores, they can be transported readily by a number of agents: flies; droplets of saliva; dust; unclean hands; refuse; water. (Pure water supplies are the subject of XI.)

(i) *Flies.* Flies are potentially excellent carriers of bacteria for three main reasons:

I. The bodies are hairy, so hold bacteria on their surface.

II. They lay their eggs on organic matter which will ferment to produce heat necessary for the development of the egg. Such organic matter will include forms of refuse and faeces.

III. The adults feed on human food.

Flies are potentially excellent carriers of bacteria which invade the intestine, *e.g.* typhoid and cholera.

(ii) *Droplets of saliva.* When a person coughs or sneezes he blows out a fine spray of saliva which will carry any organisms invading the respiratory passages. Pulmonary tuberculosis, coughs and colds (which are viruses) are transmitted in this way. Handkerchiefs should always be used when coughing or sneezing, and these should be sterilised by boiling during washing.

(iii) *Dust.* Bacteria can become attached to dust particles, and be blown about with them. Dust, being breathed in with the air, can therefore be the carrier of disease organisms. Cleanliness in the home and place of work helps to reduce infection from this source. Breathing through the nose, in contrast to the mouth, traps dust in the mucus of the nasal passages and helps to reduce infection.

(iv) *Unclean hands.* Because food is handled, it is important that the hands are kept clean, to prevent the bacteria on them from contaminating the food. It is particularly important that the hands are washed after using a lavatory. Finger-nails should be kept short and clean, as dirt harboured under them is a potential source of infection. Washing with clean water greatly reduces the number of bacteria on the surface of the skin. This number is reduced further if soap is used as well. Towels should be kept clean. Disposable paper towels are more hygienic in public wash-rooms than the roller type which are changed only periodically.

(v) *Disposal of refuse.* Refuse includes dry refuse and sewage. Both have to be disposed of effectively as they can contain disease organisms. There is also the consideration that the vast amount of dry refuse produced in houses and factories has to be removed from the premises. The disposal of sewage has been discussed already.

Dry refuse includes all the rubbish thrown away by a household or factory, *e.g.* paper, tins, glass, and vegetable matter. This is stored for short periods in dustbins or other containers and then collected by the local authority.

After the refuse is collected, it is sorted to remove any valuable material, *e.g.* metals and paper. The remaining material is either burnt in large incinerators or tipped. Tipping is rather unsatisfactory, as the tips can be breeding grounds for rats.

Dustbins should be:

I. *Kept dry*, to minimise the chances of bacteria multiplying in them. Any wet material should be wrapped in paper.

II. *Clean*. They should be washed out periodically with disinfectant.

III. *Strong*, made of galvanised iron or tough plastic. If galvanised iron is used, they should be raised off the ground to prevent them from rusting.

IV. *Tightly closed*. This prevents flies getting into them; prevents them being a food source for rats; and prevents the contents from getting wet.

(b) *Tuberculosis* (*T.B.*, *consumption*). Tuberculosis is caused by *Mycobacterium tuberculosis*. There are two varieties, one of which is found in cattle, and one only in man. Both can attack man. One of the symptoms of the disease is the presence of small lesions or tubercles in the tissue. Tuberculosis can affect a wide variety of tissues, but those most affected are the lungs and bones. The bacterium causes a gradual breakdown of the tissues, so that one of the symptoms of tuberculosis of the lungs is a cough, which ultimately develops into a spitting of blood. This is accompanied by a loss of weight, general debility, and ultimately, death.

The disease is spread by droplet infection, and through the milk of infected cows.

Tuberculosis is an endemic disease, occurring throughout the world. It has been known for centuries, but its cause was not elucidated until the bacterium was identified by Robert Koch in 1882. This is a social disease. A high incidence of tuberculosis is associated with malnutrition and poor housing conditions. Many people become infected by the bacterium, but if they are well fed, they build up a resistance to it. Malnutrition lowers the resistance and allows the bacterium to take a hold.

In 1943 mass-radiography was introduced in the U.K. enabling large groups of people, *e.g.* a school, to be X-rayed quickly and effectively. In this way lung tuberculosis is diagnosed in its early stages and treated. More recently the *Mantoux test* was introduced. The toxin of the tubercle bacillus is introduced under the skin. If the person develops a positive reaction, it means that they have been infected by the bacterium at some time. They are then X-rayed to see if the bacillus is still active. If the test is negative, they are vaccinated against tuberculosis with the B.C.G. (Bacillus Calmette–Guerin) vaccine, named after the two Frenchmen who first prepared it.

Antibiotics, *e.g.* streptomycin, viomycin, seromycin and pyrazinamide are very effective in treatment. Rest and good feeding in a sanatorium help to build up a natural resistance. In severe cases surgery may be necessary.

(c) *Diphtheria.* Diphtheria is caused by *Corynebacterium diphtheriae.* The bacterium does not penetrate deeply into the tissues, so that the symptoms are produced by bacterial toxins.

The disease is very dangerous, and very infectious. People of all ages can be infected, but the chief victims are young children. It is a notifiable disease.

Infection is by droplets or from the bedding, etc. of an infected person. The bacterium is confined to the throat, causing a sore throat and swelling of the neck glands. A white membrane develops over the tonsils and may extend further into the respiratory tract causing difficulty in breathing.

The disease is characterised by a high temperature, a

general pallor of the skin, and aching limbs. Extensive production of toxins by the bacteria cause a general poisoning of the body (toxaemia) which may affect the heart and kidneys, causing death.

The incubation period is 2–10 days, and a mild attack may last 4–6 weeks. It is possible for a person to be a carrier.

In the U.K. young children are immunised against diphtheria by giving three inoculations of the toxin at monthly intervals, when they are 2–3 months old. This preventive measure has virtually eliminated the disease in this country.

Treatment involves keeping the patient in bed, giving him plenty of fluids and injecting with the antitoxin as soon as possible after the symptoms appear. Cure is not considered effective until three cultures of swabs taken from the throat have been shown to be free of the bacterium.

(d) *Scarlet fever* (*scarletina*). This is caused by *Streptococcus pyogenes*. There are several varieties of this disease. It is spread by droplet infection and by milk. The disease used to be very serious, with ear infections, kidney inflammation, and possibly pneumonia, resulting as complications. Of recent years the bacterium seems to have lost its virulence.

The symptoms are fever, sore throat, swollen neck glands, and a rash appearing during the first 2 days. The rash begins on the face, and spreads over the body, finally affecting the limbs. It is a general reddening of the skin with small, discrete, darker red spots. The skin peels. The tongue becomes coated, and swallowing is difficult.

The patient should be isolated in bed, and should remain so for 4 weeks. He should stay in bed a week after the temperature has returned to normal. Any handkerchiefs, cotton wool, etc. which have been used in treating the patient should be burnt.

Penicillin or scarlet fever antitoxin can be used to reduce throat infection.

Scarlet fever usually affects children 1–14 years old. If possible, contacts should be quarantined for 10 days.

(e) *Typhoid fever*. Typhoid fever is caused by *Salmonella typhi* (*Eberthella typhi*). This is an intestinal bacterium which enters the body through the gut wall. The disease is characterised by high fever and violent diarrhoea. It can

the site of infection. This is frequently inside the sex organs and may appear 10–90 days after infection. Transmission to another person is particularly easy at this stage as the sore contains the spirochaetes. The second stage develops after 2–3 weeks. This is a rash, sore throat and fever. Ulcers develop in the mouth, and at this stage the disease may be transmitted by kissing. Both primary and secondary stages pass without treatment, but the disease is not cured. The third stage may take as long as 20 years to develop. Heart, blood vessels and brain are affected, and the final result is insanity.

Both venereal diseases can be cured by treatment with antibiotics (Chloramphenicol, penicillin G and tetracyclines), and gonorrhoea can be treated with sulphonamides. Early treatment is essential. People who have contracted these diseases often feel ashamed, or frightened of seeking the advice of a doctor. Such advice is essential to cure, and the consequences of allowing these diseases to develop are so terrible that it is highly advisable to overcome any personal feelings to obtain professional treatment. The body does not build up an immunity to infection by the venereal disease organisms.

VIRUSES

6. Introduction. Viruses are extremely small particles, from 300 to 30 nm in diameter. They can pass through a bacterial filter and can be crystallised out from liquids. Viruses cannot exist for any length of time outside living cells, so in this sense they are obligate parasites. They contain no metabolic enzymes, and only one type of nucleic acid. There is no nucleus and cytoplasm as in an ordinary cell, but the nucleic acid they contain is sufficient for them to use the metabolic products of the host cell to reproduce. Viruses are extremely widespread in plants and animals. In man they cause such diseases as mumps, measles, German measles, influenza, the common cold, chicken pox, smallpox and yellow fever.

Once a virus has entered the cell, it takes over the metabolic processes completely. In doing so it utilises the substance of the cell to produce new virus particles, which are released by rupturing the cell wall. The mechanism of release varies in detail.

Viruses behave as foreign protein in the blood stream, so antibodies are produced against them. Also, vaccination can be used for protection. A second line of defence is a substance called *interferon* which is produced in cells invaded by a virus. One of the difficulties in establishing immunity against viruses is their ability to produce new strains, against which the body has produced no antibodies. The various types of influenza are an example of this. Although one may be vaccinated against influenza, if the vaccine does not protect against a particular strain, infection by that strain is still possible.

7. Some examples of virus diseases.

(a) *The common cold.* This virus exists as several strains. It attacks people of all ages and is spread by droplet infection. The incubation period is about 2 days. The initial symptoms are a slight fever, secretions from the nose, and possibly a sore throat and headache. Later the temperature drops below normal. After the cold is cured, secondary symptoms of a "blocked nose" and catarrh may develop due to the secondary invasion by bacteria. Although antibodies are produced against the virus, they have a very short duration. This, combined with the occurrence of several strains of the virus, make it possible to be infected frequently by the common cold.

Colds occur more frequently in cold weather, or when one is a little "run down". There is no cure as such. The available medicines, *e.g.* throat pastilles, aspirin, help to relieve the symptoms, but do not cure the complaint. Vitamins, especially vitamin C, increase resistance to infection. Spread can be reduced by covering the mouth and nose when coughing and sneezing, and by avoiding crowded places.

(b) *Smallpox.* Smallpox is a serious disease which may reach epidemic proportions. The virus is contracted by inhalation. It attacks people of all ages, and has an incubation period of 12–14 days. The symptoms are fever and pustules appearing on the skin. The pustules dry out to leave a scab, which when it falls off, leaves a scar. Patients should be isolated until all the scabs are gone, and contacts should be traced and isolated for 3 weeks. The pustules, scabs etc. are highly infectious. This disease is notifiable in the U.K. It is rare in the U.K. as vaccination of babies against it is compulsory.

Historically, smallpox is interesting as it was the first disease against which vaccination was used as a preventive measure. In 1796 Edward Jenner immunised a small boy by vaccinating him with serum containing the related cow pox virus.

(c) *Yellow fever.* This is an interesting disease as it has mosquitoes as its vectors. The disease is confined to the tropics, and can be very serious. It was the alarming death-rate from yellow fever (and malaria) which nearly prevented the building of the Panama Canal. The natural host is a monkey. In America, mosquitoes (*Haemogogus*) feed on infected monkeys, and then on man; the man becomes infected. Another mosquito (*Aëdes*) transmits the virus from man to man. In Africa the relationship is similar, but a species of *Aëdes* transmits from monkey to man.

Inoculation against yellow fever is compulsory for people entering a country in which it is prevalent.

FUNGI

8. Introduction. The fungi are like the bacteria in that they do not contain chlorophyll and therefore live by digesting organic matter to produce new cells and energy. Like the bacteria, they cause decay, and some diseases, and are used in some important economic processes. The cells have at least one nucleus, and the fungus body may be a single cell, *e.g.* the yeasts, but more commonly made up of fine filaments (*hyphae*), *e.g.* bread mould, *Penicillium*. The mushrooms and toadstools are fungi in which the hyphae are grown into a particular shape to form a fruit-body. All the fungi reproduce by spores.

9. Beneficial fungi.

(a) *Edible fungi.* Several of the mushroom-like fungi are edible, but expert advice should be sought before eating any fungi. Some of the poisonous ones can prove fatal.

(b) *Fermentation.* The yeasts (*Saccharomyces* spp.) are

used to ferment various sugars to produce alcoholic drinks,
e.g. wines, beers, toddy. The yeast respires anaerobically
using the sugar as a source of energy. This produces ethanol
(alcohol) and carbon dioxide. Wines are made by fermenting
the sugar in grapes, using the yeast present on the grape
skin. Beers, lagers, etc. are produced by the fermentation of
maltose, produced when barley seeds germinate. Other
grains are used as sources of carbohydrate to produce other
beverages. Spirits are made by distilling the alcohol made
during these fermentations.

Citric acid is produced commercially by growing the
fungus *Aspergillus niger* in molasses.

(*c*) *Antibiotics*. Several antibiotics are produced industrially by culturing various fungi in special media. Antibiotics are substances which will kill bacteria but have no
harmful effects on the person taking them. The first antibiotic to be produced commercially with any success was
penicillin, produced in the medium by *Penicillium notatum*.

If antibiotics are used too widely in the treatment of
minor complaints, there is a danger that resistant strains of
other, more serious disease organisms develop, against which
the antibiotics would be useless.

(*d*) *Vitamins*. Riboflavin is produced commercially by
growing the fungus *Ashbya gossypii* in liquid poured off
maize soaked in water.

10. Diseases caused by fungi.

(*a*) *Diseases of plants*. Fungi cause a vast amount of
damage to food crops, *e.g.* potato blight, wheat rust.
Millions of pounds are spent annually throughout the world
on trying to control these fungi. There is not a crop which
is grown to supply man's needs which is not susceptible to a
fungus disease. As well as the damage to crops, fungi also
cause spoilage of the crop after harvesting by making it
mouldy.

(*b*) *Diseases of man*.

(*i*) *Athlete's foot*. This fungus affects skin which is regularly moist. The feet are therefore particularly susceptible.
The infection appears initially as small blisters on the soles
of the feet, and the skin between the toes exudes moisture and
peels, leaving reddened areas. If treated in the early stages,

the disease is not serious, but if left unattended, may spread over the whole foot and secondary bacterial infections may become established. Once established, the disease may persist for several years. Athlete's foot is contagious. It may be contracted in public swimming baths, gymnasia, etc., or any place where people walk bare-footed. Treatment with the antibiotic griseofulvin as a cream or powder is usually effective.

(*ii*) *Ringworm.* This fungus usually affects children before the age of puberty. There are several strains of the fungus which invade different parts of the body. As the disease is spread by contact, the form attacking the head is most common. Scalp ringworm also affects cats and dogs, so children can become infected by playing with them. The fungus grows on the skin and hair, causing the hair to fall out, and leaving bald patches. The antibiotic griseofulvin is usually an effective treatment, but treatment may have to be prolonged. It is possible for a person to become infected with this fungus without showing the symptoms. This makes it difficult to identify carriers.

(*iii*) *Thrush.* This is caused by a yeast-like fungus. It is primarily a disease of the mucous membranes of the mouth and throat on which the fungus forms a layer of white budding cells. This disease which affects young children was much more prevalent than it is now. The fungus has different strains, one of which invades the lungs, producing symptoms like those of tuberculosis. Thrush usually attacks people who are already in a poor state of health or suffering from malnutrition.

MAN IN THE LIFE CYCLE OF OTHER PARASITES

11. Introduction. Several human diseases are caused by animal parasites. In many of these cases, man is an alternative host of the parasite, the other host being some other animal. The life cycles of the parasites are usually complex, consisting of a stage in which sexual reproduction takes place, and one or more phases of asexual reproduction. This type of reproductive cycle is essential to the survival of the parasites for two reasons:

(*a*) The chances of male and female coming together are small if the parasites are enclosed separately in the host's body.

(b) The chances of an individual parasite surviving is small as its survival depends on its being able to invade a particular host species. The loss of individuals during the transfer from one host to the next is great, and is compensated for by the large number of individuals which can be produced asexually.

12. Scientific basis of control. Because the parasites are internal in man, they are very difficult to control by drugs. Drugs which will kill the parasite frequently have adverse side-effects on the patient. A great deal of research has been carried out on this problem, and the situation is improving rapidly.

Drug treatment is only effective for the individual patient, and many diseases involving alternative hosts are widespread in regions where the environmental conditions are suitable for the alternative host. Therefore the best methods of widespread and long-term control are those which eliminate the alternative host. This can be done in two ways:

(a) *Altering the environment* so that the alternative host cannot breed;

(b) *using chemicals*, *e.g.* insecticides. The use of chemicals can be very effective in the short term, but their long-term use may cause pollution problems, *e.g.* the extensive use of DDT against insects has polluted rivers and killed fish.

13. Malaria. Malaria is caused by Protozoa of the genus *Plasmodium*. It is transmitted by the *female Anopheles* mosquito. The female mosquito feeds on human blood which it needs to be able to lay its eggs. Feeding is through the fine piercing mouthparts.

(a) *Life cycle of Plasmodium.* The slender *sporozoites* are injected into the blood stream of a man with the saliva of the mosquito. From the blood they migrate to the liver where they multiply rapidly for about 12 days. During this stage there are no parasites in the blood. It is difficult to get rid of the parasites from the liver, so that once a person is infected with the parasites, the fever may reoccur for several years without re-infection by a mosquito. After the 12 days, the *Plasmodium* leaves the liver and invades the red blood cells. They are now called *merozoites*. The merozoites divide asexually in the red blood cells and break out from them to invade other red blood cells. When the merozoites

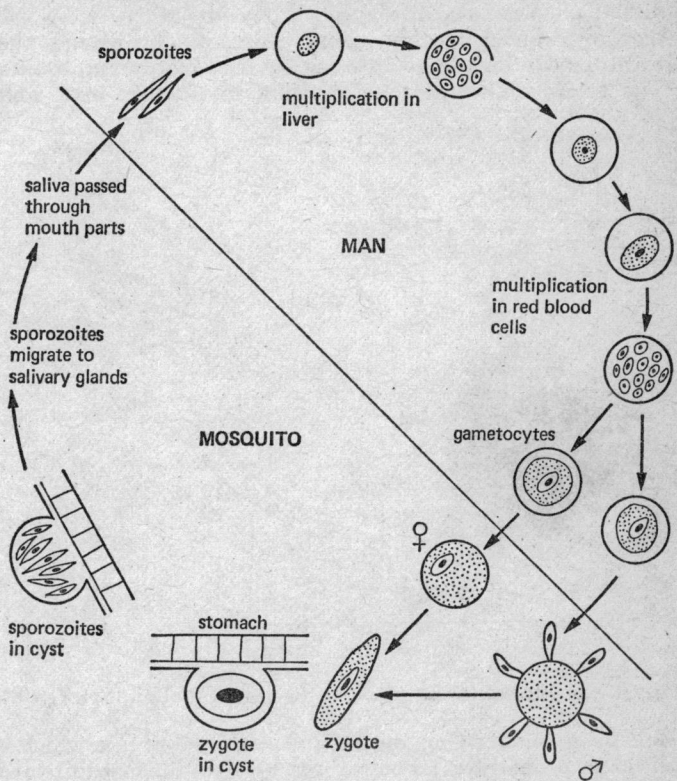

FIG. 67. Life history of plasmodium.

are released from the red blood cells, their metabolic products are also released. It is these that cause the fever. Different species of *Plasmodium* have different duration to this cycle. *P. vivax* completes the cycle once in 48 hours so that fever occurs every other day (tertian fever), while

P. malariae has a 72-hour cycle, causing fever every third day (quartan fever).

Some of the plasmodia in the red blood cells are larger than the others and are the *gametocytes* which initiate the sexual phase of the life cycle. They are taken in by the mosquito during feeding. They are cooled on leaving the warm-blooded man and entering the cold-blooded mosquito. The cooling initiates their division to produce male and

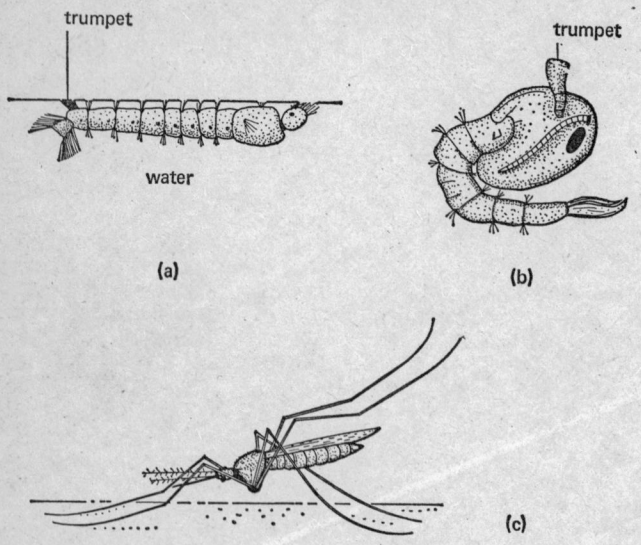

FIG. 68. Anopheles: (*a*) Larva at water surface; (*b*) pupa; (*c*) adult.

female gametes. The gametes fuse, and the zygote which is produced penetrates through the gut wall of the mosquito and produces a cyst on the outside wall. Within the cyst, the zygote divides rapidly to form a large number of sporozoites. These enter the blood system of the mosquito when the cyst bursts, and eventually arrive in the salivary glands from whence they can be transmitted to another human host (*see* Fig. 67).

(*b*) *Life history of the mosquito.* Mosquitoes are members of the Diptera (the same family which includes the flies). The female feeds on blood, while the male is relatively short-lived and does not feed. The eggs, which are laid in still water, have small air sacs on their sides which enable them to float. When the larvae hatch from the eggs they feed on small aquatic organisms. Although they live in water, they have to breathe air which they do through a small tube called the *trumpet*. This is attached to the surface of the water by surface tension. The pupa, which forms from the larva, also breathes through a trumpet. The pupa does not feed. Finally the skin of the pupa splits and the adult emerges (*see* Fig. 68).

(*c*) *Prevention and control of malaria.* There are three possible approaches to the control of malaria:

(*i*) *Preventing the alternative host* (*the mosquito*) *from completing its life cycle.* The weak links in the life cycle are the aquatic stages. Draining swamps and other stagnant water removes potential breeding grounds. Spraying the surface of the water with oil prevents the larvae and pupae from breathing. Spraying the surface of the water with the insecticide DDT is very effective in controlling the mosquito, but its residual effect on human beings is not known (it certainly kills fish). Its use should be discouraged if any other method of control is practicable. Stocking reservoirs with fish serves the twofold purpose of killing the larvae and pupae on which the fish feed, and providing the people with a valuable source of protein food.

(*ii*) *Preventing the mosquitos from biting.* This can be done by sleeping under mosquito nets, or screened rooms; and wearing protective clothing, *e.g.* long trousers and sleeved shirts. *Anopheles* begins to feed around dusk, so that protective clothing is particularly important in the evenings.

(*iii*) *Destroying the parasite within the human host.* Taking regular, small doses of antimalarial drugs greatly reduces the chances of the parasite developing, while larger doses kill it. Daraprim, Nivoquin, and Paludrine are three very effective anti-malarial drugs.

14. Sleeping sickness. Sleeping sickness is caused by a protozoon, *Trypanosoma*. There are two species of *Trypanosoma* which cause the disease in different parts of Africa. The parasite multiplies in the blood. It ultimately affects

the central nervous system causing drowsiness, which pro-
gressively develops into a coma, resulting in death. *Trypanosoma* is transmitted by the biting fly *Glossina* which

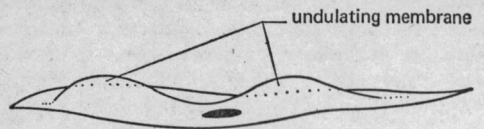

FIG. 69. Trypanosoma.

breeds in scrub land. One of the methods of control is to destroy the scrub to prevent the fly from breeding. Drugs have
been developed to combat sleeping sickness in man. A related
species causes sleeping sickness in cattle (*see* Fig. 69).

15. Schistosomiasis (Bilharzia). *Schistosoma* is a fluke which
lives in the blood vessels of human beings. It is unusual among
the flukes in that the sexes are separate. Once a male encounters

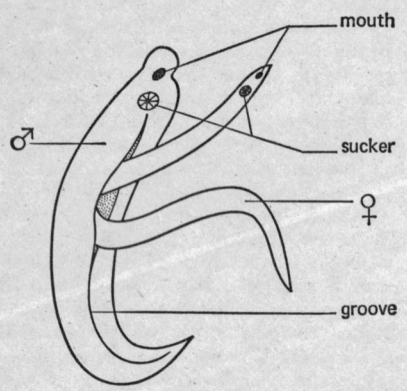

FIG. 70. Schistosoma.

a female in the blood vessel, it wraps itself around her, and
they remain permanently in contact. These flukes are about
2 cm long. When the eggs are ready to lay, the fluke migrates
to the blood vessels of the bladder (or rectum). The eggs have

a spine on them which enables them to penetrate the bladder wall, from whence they are voided to the exterior (*see* Fig. 70).

The next stage in the life cycle takes place in water snails. In the water the egg hatches into a larva which penetrates through the body wall of the snail, and rapidly multiplies asexually within the tissues. A new form of larvae (*cercaria*) is formed and released from the snail. The cercaria penetrates through the skin of a man and develops into an adult in the blood vessels. If the cercaria does not come in contact with a human host within 48 hours of being released from the snail, it dies.

Schistosomiasis can be controlled by:

(*a*) Proper sanitation in disposing of urine and faeces;

(*b*) avoiding water likely to contain the snails;

(*c*) treating the water with chemicals, *e.g.* copper sulphate, which kill the snails;

(*d*) allowing water for drinking and washing to stand for 48 hours before use. This ensures that any cercariae are dead.

This disease is widespread in the tropics, where whole local populations may be infected.

16. Tapeworm. Tapeworms live in the intestine, feeding on digested food. They are attached to the gut wall by hooks and suckers on the *scolex*. Below the scolex is a region of growth which cuts off successive segments (*proglottids*). Each proglottid contains both male and female reproductive organs. Fertilisation is internal and the eggs begin to develop soon afterwards. When the eggs are mature, all the other organs of the proglottid have degenerated, so that the proglottid is merely a sac containing eggs. These gravid proglottids are at the end of the tapeworm. They drop off and are voided to the exterior in the faeces of the host.

The embryo can survive for a long time, and when it is swallowed by the second host, it bores through the intestinal wall and migrates into the muscles where it develops into a *bladderworm*. This is a thin-walled sac containing an inverted scolex. The bladderworm remains dormant in the muscle for a long time, and when the flesh of the secondary host is eaten by a man, the muscle is digested releasing the bladderworm.

The bladder degenerates releasing the scolex. This inverts and becomes attached to the gut wall.

Human tapeworms may be over 1 m long.

There are two human tapeworms of major importance, *Taenia solium* (secondary host, the pig) and *T. saginata* (secondary host, cattle).

Tapeworms can be eradicated by:

(*a*) Proper sanitation;
(*b*) inspection of meat for human consumption, to check for the presence of bladderworm;
(*c*) making sure meat is properly cooked. This will kill the bladderworms.

17. Elephantiasis. This is caused by the nematode worm *Wuchereria bancroftii*. The adult worms live in the lymph vessels which they can block. Such blockage causes the affected part, usually the legs or scrotum, to become greatly distended. The larval worms (microfilariae) are present in the blood stream and are transmitted from one person to the next by the mosquito *Culex fatigans*.

18. Guinea worm. This nematode (*Dracunulus medinensis*) lives beneath the skin, causing severe inflammation. The males are small, but the females can be about 2 m long. They usually lie coiled beneath the skin of the ankles. When the eggs within the female are mature, the genital aperture is exposed through a small sore on the host's skin. If the sore is placed in water, the larvae are extruded in large numbers. The larvae invade the small water shrimp *Cyclops* and develop to sexual maturity. Infection of man is brought about by drinking water containing *Cyclops*.

19. Hookworms. These are two nematodes, *Ancylostoma duodenale* and *Necator americana*. Both species are widespread in the tropics. They are intestinal parasites, the adults attaching themselves to the wall of the gut by small hooks. Although they are small (about 1 cm long), the gut of an infected person usually contains a large number of them. This can be very debilitating, especially if the gut wall is damaged. The eggs are passed out with the faeces, and the larvae which develop from them enter a new host through the skin of the feet, migrating through the tissues to the gut. Warm conditions favour the survival of the larvae. Good sanitation is necessary to prevent infection by hookworms. These worms have no secondary hosts.

20. Roundworms. The roundworm is another intestinal nematode, *Ascaris lumbricoides*. It is widespread, and common where standards of sanitation are low. The female, in the intestine, produces a large number of eggs (about 100,000 a day) over several months. The larvae penetrate the intestinal wall and are carried in the blood stream to the lungs. Here they set up an irritation which causes coughing, bringing the larvae up into the mouth. They are swallowed and enter the intestine for a second time when they develop into adults. The damage to the lungs may lead to pneumonia, and possibly tuberculosis, developing as secondary infections. There are no secondary hosts of the worm.

THE HOUSE FLY (AND RELATED SPECIES)

21. Introduction. Although flies do not play any part in the life-cycle of disease-causing organisms, they are one of the primary vectors of intestinal parasites, especially the bacteria of typhoid, cholera and dysentery, and the amoeba which causes amoebic dysentery. The main offenders are the house fly (*Musca*), the meat fly (*Sarcophaga*), the bluebottles (*Calliphora*) and fruit flies (*Drosophila*).

Flies feed on solid food by covering the surface with digestive enzymes secreted through the mouthparts, and then sucking up the digested food through the proboscis. This liquid may be regurgitated as vomit spots. The food is also contaminated by the flies' excreta and by organisms carried on the hairs on the body and legs. As flies visit human faeces, manure heaps and rotting organic matter in general, their mode of feeding and the hairs on the surface make them ideal vectors for intestinal organisms.

22. Life history of the house fly. (The life histories of the other flies are similar.) The eggs are laid in small groups in rotting organic matter which provides the necessary warmth

and humidity for their development. They hatch in about
8 hours to produce white legless larvae (maggots). These feed
on the organic matter as they burrow into it. They moult
twice, finally forming a pupa inside the last larval skin
(puparium). Within this the fly develops into an adult,
emerging from the pupa after 3 days. All flies have two wings,
in contrast to other winged insects, which have four.

23. Control of fly-disseminated diseases.

(*a*) Food should be covered.
(*b*) Manure and compost should be kept far from houses.
(*c*) Proper sanitation.
(*d*) Dustbins should be covered.
(*e*) Town refuse should be burnt.
(*f*) Insecticidal sprays can be used, but only with
caution. Insecticides on foods may have effects on human
beings as yet unknown.

CANCER

24. Introduction. A cancer develops when some of the body
cells begin to divide abnormally. When they do they may invade
adjacent tissues, or may even migrate through the blood and
lymph vessels into other parts of the body. There is no single
disease called cancer, and different cancers may be caused by
different agencies.

Cancers are recognised by:

(*a*) New growth in a tissue or organ;
(*b*) the cells of old and new growth being different;
(*c*) the appearance of unusual cells in organs.

Cancers can affect almost any organ of the body, but most
affect the blood (leukaemia), the lungs, the stomach and large
intestine, the uterus and the breasts.

25. Causes of cancer. There may well be no one cause of
cancer, or different cancers may be caused by different agencies, *e.g.*:

(*a*) *Viruses.* There is a great deal of evidence to suggest
that some cancers are caused by viruses. Virus-like particles

capable of producing the disease symptoms have been isolated from various cancers, *e.g.* mouse leukaemia.

(*b*) *Somatic mutations*. Cells in the tissues may mutate and begin to divide out of phase with the cells of the tissue.

(*c*) *Inheritance*. Some forms of cancer occur more frequently in some families than in others. This suggests that they may be inherited.

(*d*) *Carcinogens*. A wide range of chemicals have been shown to induce cancers. These are called carcinogens.

(*e*) *Radiations*, *e.g.* X-rays have been shown to induce cancers.

Whether radiations and carcinogens actually cause cancer is difficult to determine. They may simply activate the virus, or cause the mutation which results in the cancer.

26. Treatment of cancer.

(*a*) *Surgery*. If a cancer is a confined tumour, it is possible to remove it by surgery. The difficulty is to make sure that all the cancer cells are removed.

(*b*) *Radiation*. γ rays (from radioactive cobalt or caesium) or X-rays are focussed onto the cancerous cells and destroy them.

(*c*) *Chemotherapy*. A great deal of research is being carried out to find drugs which will kill cancer cells.

27. Lung cancer. A great deal of attention has been focussed on lung cancer during recent years. Between 1944 and 1955 the number of deaths from lung cancer in the U.K. more than doubled (from 6,568 to 17,271). This dramatic increase is correlated with an increase in tobacco smoking. Tobacco smoke contains carcinogenic substances, and there is a correlation between the number of deaths from lung cancer and the number of cigarettes smoked. That is, a heavy cigarette smoker (more than 20 a day) who inhales the smoke is more likely to contract lung cancer than a non-smoker, or pipe smoker.

Although this connection between cigarette smoking and lung cancer does exist, it does not follow that smoking causes lung cancer. Other pollutants of the atmosphere also cause lung cancer. These are statements of fact, but do not pretend to state the cause. For example, it is possible that people who are already infected with the virus of cancer are the type who feel the need to smoke tobacco.

Whether smoking directly causes lung cancer or not, a smoker's lungs become coated with various tars from the smoke. He is more susceptible to bronchitis, and his breathing is impaired. In the light of these facts, the government of the U.K. have thought it advisable to print a health warning on cigarette packets and to prohibit the advertising of cigarettes on television. Whatever the scientific arguments, the risk of contracting lung cancer is not worth taking, and one would be strongly advised not to start smoking, or if one begins smoking, to give it up.

PROGRESS TEST 10

1. Give two ways in which bacteria differ from true cells. (3)
2. Give two ways by which bacteria are identified. (3)
3. State three ways in which bacteria reproduce. (3)
4. State seven ways in which bacteria are useful, or are used by man. (4)
5. State six ways in which bacteria are spread. (5)
6. Name eight diseases caused by bacteria. (5)
7. Name eight diseases caused by viruses. (6)
8. How does the body protect itself from virus infections? (7)
9. State four ways in which fungi are beneficial to man. (9)
10. Name three diseases caused by fungi. (10)
11. Name seven diseases caused by parasitic animals. (11-20)
12. Name the vector of malaria. (13)
13. Name the vector of sleeping sickness. (14)
14. How are cancers recognised? (24)
15. Give five possible causes of cancers. (25)
16. Give three possible treatments of cancer. (26)

EXAMINATION QUESTIONS

1. Write an essay to explain the statement that man is the great consumer of the earth's resources.
2. Explain the parts played by bacteria in: (a) the carbon cycle; and (b) the nitrogen cycle.
3. Describe how a sewage farm functions, paying special attention to the part played by bacteria in the process.
4. Give an account of the spread of bacteria under the following

headings: (*a*) water; (*b*) food; (*c*) flies. How can their spread be prevented?

5. Why is the efficient disposal of refuse important? Outline the various methods of refuse disposal.

6. Name three water-borne bacterial diseases. Describe the symptoms and control of *one* of them.

7. Tuberculosis is considered to be a social disease. Explain why it was more common 100 years ago than it is now. What factors have brought about the decrease in the incidence of the disease?

8. What are the causes of the two venereal diseases? Describe the symptoms of syphilis. How are venereal diseases best prevented and treated?

9. Outline the principles of control of animals parasitic on man.

10. Give an account of the life history of a malarial parasite. Outline the methods of controlling malaria.

11. Write a brief account of diseases of man caused by nematodes.

12. Discuss the evidence that lung cancer is caused by cigarette smoking. What other possible causes are there for this disease?

CHAPTER XI

WATER

NEED FOR WATER

1. Personal needs. The amount of water required by the modern household is remarkably large. Each person needs it for:

    (*a*) Drinking;
    (*b*) washing;
    (*c*) flushing toilets.

2. Household needs. The whole household needs water for:

    (*a*) Washing clothes;
    (*b*) washing dishes;
    (*c*) cleaning the house;
    (*d*) various extras, *e.g.* washing the car, and watering the garden.

The volume of water required for each of these processes is going to vary from household to household, and from season to season.

As well as the domestic requirements, the water demands of industry are enormous, especially as a coolant.

THE WATER CYCLE AND WATER SUPPLY

3. The water cycle. All water originates as clouds which are masses of water vapour. Fine water droplets in the clouds fall as rain, snow, hail and dew. The total water landing on a surface is called the *precipitation*. The precipitation may:

    (*a*) Land directly in a large water mass, *e.g.* the sea or lakes;
    (*b*) Land on the earth's surface and percolate through it to emerge as springs or wells which flow into larger rivers and hence to the sea;

(*c*) Remain for a relatively short period on the surface of the earth, *e.g.* dew or rain after a shower. In any case, some of this water is evaporated by the heat of the sun. Clouds are formed and the process is repeated.

Water is added into the cycle by such natural processes as respiration, and by burning hydrocarbon fuels. Water is removed by photosynthesis (*see* Fig. 71).

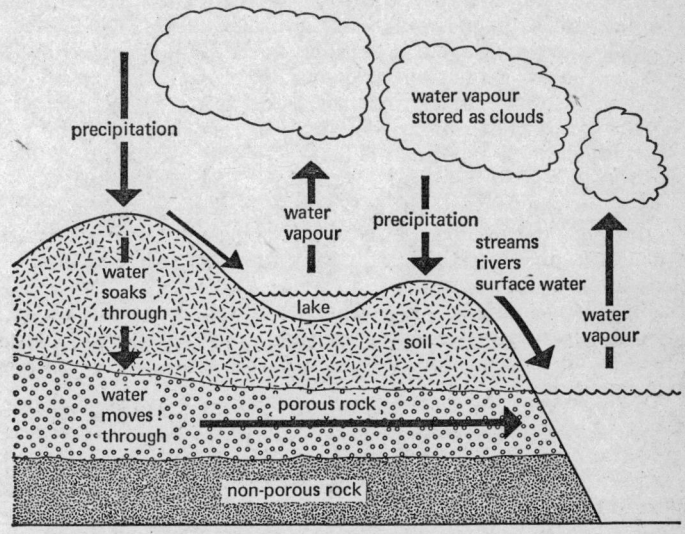

FIG. 71. Water cycle.

4. Water supply.

(*a*) A small community in an under-developed country may use a *stream* as its water supply. The risk of contamination under these conditions is high.

(*b*) *Rain*, landing on a porous rock, will soak through it until it reaches an impervious layer, *e.g.* clay. The water will run along the impervious layer, and if it reaches the surface, *e.g.* on the side of a hill, it will form a *spring*. Spring water is usually pure as it will be filtered as it runs through the rocks. It may not be suitable or pleasant to drink, because of the salts it has dissolved while passing through the rocks.

(c) *Wells* are dug through the porous rock until the impervious layer is reached. Frequently, the sides are built-up with stones or brick. They are kept filled by water draining along the impervious layer of rock, forming an underground reservoir. If the edges of the drainage basin are higher than the centre, the water at the centre will be under pressure, so that it will be forced to the surface of any well sunk in the area. This is an *artesian well.*

(d) *Reservoirs.* The rapid increase in population, coupled with the ever-increasing demands of industry, made it obvious to the local authorities of the early nineteenth century that naturally occurring sources of water would not be sufficient to meet the demands of the community. This initiated the building of reservoirs. The first was built in 1834 by the Metropolitan Water Board to supply parts of London. It was made at Stoke Newington. Reservoirs are usually constructed in high rainfall areas, or by damming rivers.

(e) *Desalination.* (The removal of salt from sea water.) In low rainfall areas this method of producing drinking water and water for irrigation is becoming an economic proposition.

PURIFICATION OF WATER

5. Water consumption. Spring water is usually sufficiently pure to drink without further treatment, but, if the water has been collected from some overground source, it will contain micro-organisms and organic matter. These may, or may not, be harmful, but may colour or flavour the water. Before water is used by the public it must be:

(a) Free from harmful organisms;
(b) colourless;
(c) odourless;
(d) tasteless.

6. Methods of purification. Purification is brought about:

(a) *Naturally.* When water is allowed to move slowly, as in a reservoir, a large proportion of the suspended solids sediment out. Bacteria and other micro-organisms break down the organic matter to carbon dioxide and water, while harmful bacteria are destroyed by exposure to sunlight and by the oxygen produced during photosynthesis by green plants.

(*b*) *By artificial filtration.* Before water is supplied for consumption it is passed through *filter beds* as shown in Fig. 72.

The water flows slowly into the top of the filter bed and runs down through the particles of varying size. This removes

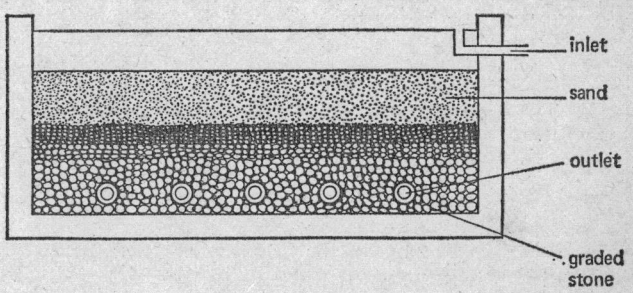

FIG. 72. Filter bed.

any solid matter that is present. A layer of algae and other micro-organisms is allowed to grow on the surface of the sand. This acts as a bacterial filter, removing harmful organisms. When this layer becomes logged it is replaced with fresh sand. A new film of algae, etc. is allowed to grow before the filter bed is operated again.

Water may be filtered through *pressure filters* which are closed concrete tanks filled with sand.

(*c*) *By chlorination.* After filtration, chlorine is added to the water at a rate of 1 part per million. This kills any harmful bacteria which may have passed through the filter, and oxidises any organic matter.

7. Water pollution. When we mention pollution in this context, we mean that the water is unfit for human consumption. The naturalist would mean that the water could not support wild life. For example, water containing a high percentage of human excreta would be polluted in our sense, but may form an excellent environment for aquatic life. The following are the main sources of pollution:

(a) Sewage;

(b) metallic ions, e.g. copper, zinc and lead, which may be from natural sources or from industrial plant;

(c) soap and other detergents which affect the efficiency of filter beds;

(d) insecticides washed off agricultural land into rivers;

(e) specific chemicals discharged from industrial plant into water.

PROGRESS TEST 11

1. Give six ways in which water is used in a household. (1, 2)

2. Draw a diagram to illustrate the water cycle. (3)

3. State five ways by which a community may obtain its water. (4)

4. What are the four criteria which must be satisfied before water is suitable for consumption? (5)

5. What are the three processes by which water is purified? (6)

6. Give five main sources of water pollution. (7)

EXAMINATION QUESTIONS

1. Outline the water cycle in nature.

2. Describe the formation of: (a) springs; (b) artesian wells. Why may this water not be suitable for drinking?

3. Describe the methods used to purify water for consumption in a town.

4. What methods can be adopted to prevent the pollution of water supplies?

CHAPTER XII

AIR POLLUTION

PRIMARY CAUSES OF AIR POLLUTION

1. Burning coal and oil. These produce smoke, dust and sulphur dioxide as pollutants. "Britain consumes annually about 2,000,000,000 tons of coal and 25,000,000 tons of oil. The output of noxious products is estimated at 1,000,000 tons of dust, 2,000,000 tons of smoke, and 5,000,000 tons of sulphur dioxide" (Kenneth Mellanby in *Pesticides and Pollution*).

2. Smoke and dust. These cover the surfaces of buildings and dirty clothing. More important, when mixed with droplets in a mist, they cause *smog* which produces respiratory troubles that can be fatal. These factors may be responsible for the greater incidence of lung cancer in towns than in the country.

3. Sulphur dioxide. Sulphur dioxide in the atmosphere tarnishes metal and can have a deleterious effect on vegetation. When dissolved in water, it forms a weakly acid solution which will accelerate the damage to buildings made of limestone or marble.

4. Carbon monoxide. This gas is produced by the incomplete combustion of petrol in car engines. The gas is extremely poisonous, and a danger level can be reached in the centres of towns where there is a large amount of traffic, and the circulation of air is limited. In Tokyo, the problem is so acute that the police have to be provided with oxygen when they are on traffic duty.

5. Lead. A compound of lead is added to petrol as an "anti-knock", making combustion more efficient. This is vapourised and discharged with the exhaust gases. Although no effects from lead poisoning from this source have been reported yet, lead may well accumulate in plants and be consumed by people and animals.

6. Fluorine. Fluorine is produced in the smoke from brick works and iron and aluminium plants. This is accumulated by plants, and may well enter the food chains which end with a human consumer. Small amounts of fluorine in the diet reduce tooth decay, but larger quantities cause fluorosis. Initially, the teeth become mottled and rough, but abnormalities of the bones are caused in more serious cases.

PREVENTION OF AIR POLLUTION

7. Legislation. A public awareness of the problem has led to legislation in many countries to restrict the production of pollutants. Such legislation includes:

(*a*) The establishment of *smokeless zones* in which the burning of conventional coal fires and the production of smoke by industry is prohibited;

(*b*) the restriction on the composition of smoke produced by factories;

(*c*) the control of the composition of the fumes produced by internal combustion engines.

8. Research. A great deal of research is being carried out on:

(*a*) Production of smokeless fuels;

(*b*) reduction of exhaust fumes from motor vehicles;

(*c*) development of other forms of engines for cars, etc.

ATOMIC RADIATION

9. Problems of pollution. Atomic radiations are produced on a large scale by the explosion of atomic bombs, but in the long term the accidental discharge from atomic power stations and industry may prove more hazardous. X-rays and other radiations used in medicine need careful control and supervision.

Radiation ionises some of the chemicals dissolved in cells. This increases their activity. If the radiation dose is too high, the cells are killed.

The *rad* is one of the more common units in which radiation is measured. The cosmic rays which are constantly bombarding the earth from space expose man to about 0·1 rad/year. This level must be safe. Exposure to 1,000 rad will kill in a matter of a few days, but much lower exposures, *e.g.* about 50 rad, can cause mutations in the cell, and if they affect the ovaries or testes, may cause sterility.

Immediate death from radiation is likely only in the vicinity of an atomic explosion. The problem lies in the possible damage caused by prolonged exposure to relatively small doses. For example, people working near radioactive sources may, over the years, show symptoms of radiation sickness. One of the major hazards is the accumulation of radioactive material in plants which ultimately enter the food cycle of man. A radioactive form of strontium (strontium 90) is produced during an atomic explosion. It enters the atmosphere and is washed into the soil by the rain. From the soil it enters plants, particularly grasses, which are eaten by cattle. Milk produced by the cattle contains strontium 90, and when this is fed to children, it enters into the composition of the bones, causing malformation.

10. Disposal of radioactive material. Radioactive materials are impossible to destroy so disposing of them presents a great problem. Dumping them in the sea is no solution because of the very small concentrations that are acceptable in the environment. The most acceptable method of disposal is to close them in a radiation-proof container, *e.g.* of lead and concrete, and dump the container in the deeper parts of the ocean.

PROGRESS TEST 12

1. State six primary causes of air pollution. **(1–6)**
2. What legislation has been enacted to reduce air pollution? **(7)**
3. What research projects are in hand to reduce air pollution? **(8)**
4. Give three problems peculiar to pollution by atomic radiation. **(9)**

EXAMINATION QUESTIONS

1. Choose any three air pollutants produced by industry. What are their effects: (a) on the environment; and (b) on individual people? How can they be reduced or eliminated?

2. "The motor car is a major user of the earth's energy resources and a prime polluter of the atmosphere." Discuss this statement.

3. Atomic power may become the world's major source of energy. Discuss the dangers involved in using atomic energy. What precautions have to be taken to reduce the hazards?

CHAPTER XIII

HOUSING

CLIMATIC CONSIDERATIONS

1. Introduction. The type of housing varies considerably with the climate in which people are living, *e.g.* heating and ventilation are not such problems in the tropics as they are in colder climates. The availability of building materials is also going to influence the type and size of houses.

2. General requirements in temperate climates.

(*a*) *Waterproof*. The roof and walls should not allow water to penetrate, neither should water soak up through the walls as damp. A roof of overlapping tiles, slates or other material is sufficiently waterproof, and a damp course of waterproof paper or plastic built into the walls near ground level prevents water rising from the soil. The floor boards of the ground floor are raised above the foundations with a cavity below them open to the exterior by ventilation bricks. This allows air to circulate under the floor boards, keeping them dry.

(*b*) *Warmth*. Heat loss is reduced by providing a layer of still air between the external environment and the rooms of the house. This is done by:

(*i*) *Cavity walls, i.e.* walls built of two layers with an air space between them. Recently the practice of filling this cavity with plastic foam is becoming increasingly popular.

(*ii*) *Covering the ceilings* of the upper rooms with fibre-glass or similar material.

(*iii*) *Double glazing*. This is a window with two panes of glass trapping still air between them.

(*c*) *Hot and cold water* should be available in the house, both in the bathroom and kitchen.

(*d*) *Kitchen sink* should be present.

(*e*) *Bathroom*.

215

(*f*) *Water closet* either separate from, or in the bathroom.

(*g*) *Cooking facilities*.

(*h*) *Food storage*.

(*i*) *Fuel store*.

(*j*) *Refuse disposal*. A space for dustbins or other method of refuse disposal should be available.

SIZE OF HOUSE

3. Overcrowding. It is considered that the minimum floor space required for two people is 10.5 m^2, with a minimum ceiling height of 2.25 m. This is a small room for two people to live, eat, sleep in throughout the day, but there are thousands of people who have to live in worse conditions than this in all the temperate countries. The problem has arisen through the increase in population over the last 150 years. The building of new houses has not kept up with this increase, and older buildings have fallen into disrepair. Governments are aware of the problem and a slum clearance programme and house-building are high on the list of necessary social improvements.

4. Multi-storey flats. Building more houses necessitates using more land; this reduces the amount of land available for agriculture, recreation, etc. Because of this difficulty, many of the new dwellings are in multi-storey flats. This type of dwelling introduces psychological problems. Living 20 storeys up in a block of flats surrounded on all sides, above and below by other people; having to go down a large number of stairs to reach the street; no convenient place for the children to play near at hand; all these go to make a completely alien environment to which many people do not adapt easily.

VENTILATION

5. Need for ventilation. In homes and places of work, it is necessary to replace the stale air. People living in confined spaces make the air stale by:

(*a*) Replacing the oxygen by carbon dioxide during respiration;

(*b*) increasing the humidity by breathing and perspiring;

(*c*) increasing the temperature.

People in a non-ventilated room begin to yawn and become lethargic due to the deterioration in the composition of the air which they are breathing.

Bad ventilation increases the chance of infection by air-borne organisms.

One of the criticisms of houses, factories and schools built during the period following the Industrial Revolution is their lack of ventilation. These buildings are being replaced by modern ones in which the minimum window and floor space available per person is stipulated by the Factory and Buildings Act 1948.

6. Methods of ventilation. Any method of ventilation depends on the fact that warm air is less dense than cold air and rises to the top of a room. Therefore, there must be openings high in the room to allow the stale air to escape, and others lower down to allow the colder fresh air to enter.

(*a*) *Fires and chimneys.* A fire lit in the hearth warms the air which rises up the chimney to be replaced by air coming from under doors, etc. This is an excellent method of ventilation but often causes unpleasant draughts. Central heating does not provide this kind of ventilation, so that centrally heated rooms have to be adequately ventilated by other means.

(*b*) *Air bricks.* These are bricks with holes through them. They are set in the wall below the floor boards to keep the floors damp-free, and high on the walls of a room to release exhausted air.

(*c*) *Windows.* There are various types of windows, but all are designed to allow air to leave rooms as well as enter them. They are placed in the upper half of the wall so that the hot, stale air can leave through them. Some have small fanlights high in them to allow the exit of air without annoying draughts.

(*d*) *Electric fans*, etc. If there is insufficient ventilation, electric fans, or similar devices, are fitted high in the walls to forcibly extract the stale air.

(*e*) *Air conditioning.* Modern buildings frequently have air conditioning plants which draw fresh air into a building, warm it to a suitable temperature (or cool it), correct the humidity, and filter it.

LIGHTING AND HEATING

7. Lighting. Adequate sunlight is needed to:

(a) *Be able to see well in the room.* Too little, or excess light can lead to eye strain.

(b) *produce vitamin D in the skin,* and have a bacteriocidal effect. Glass eliminates the ultra-violet light responsible for these effects, so windows should be open as much as possible.

(c) *produce a pleasant environment.* This is psychologically important. A light, cheerful room makes one feel cheerful, while a dull room is depressing. The decoration of a room is important in this respect, as the colour scheme combined with the amount of light can create a definite psychological atmosphere in a room.

Ideally one-third of the wall space of a room should be window.

If the light is inadequate it should be supplemented by:

(a) Electric light;
(b) gas light;
(c) paraffin (kerosene) lamps;
(d) candle light.

Electric lights are either of the filament (bulb) type, or fluorescent. Both give a good light, but the latter does not cast a shadow. Fluorescent lights are used in kitchens, factories, etc. where shadows are irritating, while filament lights are used in sitting rooms, etc., where the variety of light and shade make the room more interesting.

In urban areas, and most rural areas, electric lighting has superseded the other three types. None of these three gives such a good light as electricity, and they produce carbon dioxide as the fuel burns. This adds to the pollution of the atmosphere of the room.

8. Heating. For general comfort, houses in cooler climates have to be heated artificially. A temperature of about 10°C is suitable for corridors and bedrooms; 15–17°C in working rooms, *e.g.* classrooms; while 17–18°C is preferred in living rooms where people are sitting and relaxing. Various forms of artificial heating are used, the choice of fuel being largely dictated by the price.

(*a*) *Central heating*. There are various forms of central heating fuelled by gas, oil, or coal. Most domestic types involve the heating of water which is circulated to the various rooms where the heat is released through radiators. This type of heating has the advantage that the whole house is heated to a minimum temperature, but the temperature of the separate rooms can be varied. This is achieved by thermostats. The normal water temperature is about 82°C. This type of heating is reasonably economical, clean, and needs little attention. The disadvantages are:

(*i*) Ventilation is impaired;
(*ii*) the walls above the radiators become dirty due to dust deposited there by convection currents.

(*b*) *Coal fires*. Coal burnt in an open grate gives good ventilation to the room, causing hot air to be drawn through the room and up the chimney. Although coal is usually relatively cheap, about 80 per cent of the heat produced is lost up the chimney. Coal-burning also produces a lot of soot, which adds to the atmospheric pollution and dirties the room. Smokeless coals, while reducing atmospheric pollution, are expensive and produce a fine dust which makes the room dirty.

(*c*) *Gas*. This is one of the most popular fuels for cooking, and is becoming increasingly popular for central heating. It is clean, and when judged by its heat production is (on today's prices), the cheapest household fuel. When used for room heating, adequate ventilation, *e.g.* a flue, is necessary, so that gas fires also ventilate rooms. Gas (butane and propane) is often supplied in cylinders for use where mains gas is not available.

(*d*) *Oil*. Oil heating is relatively cheap, but oil prices are rising rapidly. Paraffin heaters are a convenient, portable form of heating but are potentially dangerous. They are a fire hazard, and should never be moved when lit. They are easily knocked over and spilt paraffin is an additional fire hazard. If oil heaters are not properly maintained they soon smell. Good ventilation is needed as they use up the oxygen in the room, producing copious amounts of carbon dioxide and water.

(e) *Electricity*. Electricity is used for heating and cooking. It is a clean source of heat, but is the most expensive. (The use of night-storage heaters can reduce the cost.) Electric heaters may be direct heaters, convectors or radiators in which an element is used to heat oil or concrete. Electricity is controlled thermostatically with ease, and is a quick form of heating.

PROGRESS TEST 13

1. State three ways used to keep houses dry. (2)
2. State four ways of preventing heat loss from buildings. (2)
3. Give three reasons why rooms should be adequately ventilated. (5)
4. Give five methods of ventilating buildings. (6)
5. Give four reasons why there should be adequate sunlight in a room. (7)
6. What temperature is most suitable for classrooms? (8)
7. What five methods can be used for heating buildings? (8)

EXAMINATION QUESTIONS

1. Explain the uses of: (a) cavity walls; (b) air bricks; (c) fan lights; (d) damp courses; in buildings.
2. If you were buying an old house, what would you do to reduce your fuel bills?
3. If you were designing a factory what would you do to ensure there was: (a) adequate lighting; (b) adequate sanitation?
4. What are the problems of housing a large number of people in a small area of land? If you were designing a new housing estate what would you do to overcome these problems?

CHAPTER XIV

MAN IN A COMMUNITY

DEVELOPMENT OF COMMUNITY RESPONSIBILITY

1. The community relation. Since prehistoric times, man has lived as a communal animal. He discovered that hunting as a group was more efficient than hunting as an individual. Agriculture is more efficient if run as a communal enterprise, with members of the community helping each other at busy times of the year, *e.g.* during harvesting. The development of this community relation throughout the ages led to the establishment of community enterprises, *e.g.* in towns and villages. In such a relationship, the individual has responsibilities to the community, and the community accepts responsibility towards the individual, *e.g.* the maintenance of the poor of the parish under the Elizabethan Poor Laws.

2. Industrialisation. As the population increased, and became industrialised, the situation altered, leading to the evolution of a working class. In the early stages of the Industrial Revolution workers were exploited so that the owners of factories, etc. could make the maximum profit. Working people lived in appalling conditions and the death rate due to malnutrition, disease and industrial accidents was high. Young children and women worked in heavy industries for very long hours for low wages, while the wages paid were such that all the members of the family had to work to provide even the minimum necessities of life. By about 1820, the situation was beginning to be rectified, and the employers gradually began to realise that people worked better if they were healthy and happy than if they were oppressed.

MODERN COMMUNITY CARE

Let us consider some of the ways in which the community cares for its individual members.

3. Factories and places of work.

(a) The minimum age at which children can be employed is laid down by law.

(b) The maximum number of hours that children can work is limited by law.

(c) Minimum safety standards are enforced.

(d) Minimum sanitary standards are laid down.

(e) There are minimum standards of lighting and heating.

(f) The length of meal-times and breaks during the day are agreed.

(g) Larger factories, shops, etc. have doctors or nurses in regular attendance.

(h) Many firms provide recreational facilities, e.g. clubs and playing fields, for their employees.

4. Education.

In 1870 an Education Act was passed in which the full-time education of children from 5 to 14 years old was made compulsory. (The upper age limit is now 16 years old.) From this time onwards, the schools have become increasingly concerned with the welfare of the pupil as well as their education. These services (in conjunction with the Health Service) include:

(a) Subsidised meals and milk (free if under 7 years old);

(b) medical check-ups;

(c) dental treatment;

(d) eye sight checks;

(e) mass radiography as a check for tuberculosis and other chest complaints.

This medical treatment is under the control of the School Medical Officer for the county or borough.

Special schools or classes are provided for mentally or physically handicapped children.

5. Medical services.

In the U.K. everyone is given free, or heavily subsidised medical treatment. To pay for this the employer and the employee each pay a weekly contribution. The Health Service includes:

(a) Treatment by a doctor;
(b) ambulance service;
(c) reduced charges for prescribed medicines from a chemist;
(d) home nursing;
(e) home helps;
(f) prenatal and antenatal clinics;
(g) attendance by a midwife;
(h) dental treatment, at reduced or no cost;
(i) consultation with an optician and spectacles at reduced, or no cost.

6. **Welfare service.** This is a wide-ranging service concerned with the general welfare, rather than the health of the community, but it is closely linked with the medical services. The service is run by trained welfare officers, some of whom are voluntary. It includes:

(a) *Child welfare service*, which looks after the needs of children deprived at home by neglect, or illness, etc. of the parents. In extreme cases, foster-parents for the children are arranged, or the children may be adopted.

(b) *Housing officers*, who check the standard of accommodation, and have legal powers to compel landlords to improve property to a minimum legal standard, or can arrange alternative accommodation if families are overcrowded.

(c) *Family planning clinics* where people are advised as to the various methods of contraception available.

(d) *Citizens' advice bureaux* which help people with legal and similar problems.

(e) *Legal aid* which enables those who cannot afford the full fee to have the services of a lawyer free, or at a fee which they can afford.

7. **Pensions and other financial benefits.**

(a) All men over the age of 65 years, and women over the age of 60 years are entitled to a state retirement pension,

even if they are still gainfully employed but providing they
fulfil certain conditions. This is paid for from the National
Insurance contribution.

(*b*) People who are normally employed and are out of
work are paid an unemployment benefit.

(*c*) A person who is unable to work through illness is paid
a sickness benefit.

(*d*) A family allowance is paid to every family with two or
more children.

(*e*) Widows are paid a state pension.

(*f*) If the family income is below a statutory limit, a
supplementary benefit is paid.

(No values are given for the charges and pensions, because they
are altered frequently.)

PROGRESS TEST 14

1. Give six ways in which working conditions are controlled
or improved. **(3)**

2. State five ways in which schools help to supervise the health
of the pupils. **(4)**

3. What two main services are provided from the National
Insurance contributions? **(5, 7)**

4. Give eight benefits available under the National Health
Service. **(5)**

5. State five services available from the welfare service. **(6)**

6. Give six financial benefits available under various State
schemes. **(7)**

EXAMINATION QUESTIONS

1. Outline the part played by legislation in: (*a*) regulating the
conditions of work; (*b*) the provision of financial benefits.

2. Describe the functions of the National Health Service: (*a*)
nationally; (*b*) regionally.

3. Describe the functioning of the welfare services. How are
these services financed?

EXAMINATION TECHNIQUE

Many boards are now setting three types of examination paper —a traditional essay paper, a structured answer paper and a multiple-choice paper.

1. The essay paper. In this paper you have a choice of questions, and each question carries the same number of marks. Therefore, it is important that your last question is as good as your first. Many candidates do not do themselves justice in examinations because they produce excellent answers to the first question, and their marks get successively less, until their last question is well below the pass mark.

To achieve the balance it is essential to allocate your time equally to each question. Naturally, the question at which you are most competent is attempted first, but frequently this means that far more is written for the answer than is necessary. The examiner abides by a mark schedule rigidly, and will not allocate marks for unnecessary or irrelevant facts. For example, consider the following question:

"Describe the digestion of food by a human being. What is the importance of (a) carbohydrates, (b) iron, and (c) vitamins in the diet?"

This question does *not* ask for a detailed description of the alimentary canal, or for a list of sources of carbohydrates, iron and vitamins. The key words here are "describe" and "importance", so that any information not directly concerned with these two points is irrelevant.

It is important to strike the right balance within the question. The question above may be marked (out of 20): digestion—8 marks; carbohydrates, iron, vitamins—4 marks each. If most of your time was spent on the digestion section, you may obtain 8 marks for that, and then possibly only 1 each on the other three sections, giving a total of 11. A more balanced answer may give 6 for digestion, but 3 for each of the other three sections—a total of 15!

2. Structured answer paper. This paper contains four or five questions each of which begins with a statement, problem or diagram on which several questions are set. For example, a question on the kidney may begin with a diagram of a section through a kidney. The questions may be (a) label structures A–D, (b) what are the functions of the structures you have labelled?, (c) state three changes in the composition of the blood passing through the kidney.

This type of question is easier to answer than the essay type, because you receive more guidance. Be as brief as possible in your answer, sometimes one word will be sufficient. The space provided for the answer on the paper is not always a guide to the length of the answer. Never try to write an essay for the answer.

As all the questions are compulsory, they do not necessarily carry the same marks. Rather you should think of the paper as a whole with one mark for each correct point.

3. Multiple-choice paper. (*See* Appendix II.) Here you have to choose the one correct answer from possible alternatives. This paper is marked by computer, so make sure that you underline only one answer to each question—more than one underlining is marked wrong. Make sure that you finish the paper; even if you are not sure of the answer to a question underline it, you may be right, and you do not lose marks for being wrong.

You have a choice of five (sometimes four) answers to each question, one of which is right. The best method of arriving at the correct answer is this. First of all, eliminate the alternatives which are obviously wrong—this will usually leave you with three to choose from. Check these three carefully, as the differences between them can be quite small. Eliminate the two wrong ones, which leaves you with the correct answer. These papers are as difficult as the others, and the quick answer after a superficial reading of the question can easily be wrong.

Questions of variable difficulty are scattered throughout the paper, so that it is better to go through the paper and answer all the questions that you can first and then go back and spend time on the more difficult ones.

4. Developing an examination technique. Here, like in any other technique, practice makes perfect. It is essential to practice examination questions of the type you will be sitting under examination conditions. For essay-type questions, prepare answers from the textbook and then answer them in the allocated time without further reference to the book. When you have completed the course attempt papers of all types under examination conditions to make sure that you can finish them in the allocated time.

MULTIPLE-CHOICE QUESTION PAPER

Questions 1–10. Select the best response in the following questions.

1. Which of the following enzymes is produced in the stomach?
   A. ptyalin   B. pepsin   C. trypsin   D. amylase
   E. lipase.
2. The outer layer of the tooth is called:
   A. pulp   B. dentine   C. enamel   D. cement   E. crown.
3. Which is *not* part of the eye?
   A. lens   B. aqueous humour   C. retina   D. cornea   E.
   eyelid.
4. Urea is produced in:
   A. liver   B. kidney   C. stomach   D. pancreas   E. ileum.
5. Sewage is rendered harmless by:
   A. gravel   B. viruses   C. sedimentation   D. bacteria   E.
   filtration.
6. The atmosphere contains which percentage of oxygen?
   A. 20   B. 2   C. 80   D. 0·03   E. 50.
7. Which of the following is a good source of vitamin C?
   A. polished rice   B. tomatoes   C. eggs   D. fish liver oil
   E. bananas.
8. Which of the following represents a reflex arc?

   A. receptor, motor neurone, spinal cord, sensory neurone,
      effector.
   B. effector, motor neurone, spinal cord, sensory neurone,
      receptor.
   C. receptor, sensory neurone, spinal cord, motor neurone,
      effector.
   D. effector, sensory neurone, spinal cord, motor neurone,
      receptor.
   E. receptor, effector, spinal cord, motor neurone, sensory
      neurone.

9. Which of the following is a bone of the arm?
   A. tibia   B. fibula   C. femur   D. pelvis   E. radius.
10. Which of the following is an antibiotic?
    A. aspirin   B. streptomycin   C. heroin   D. iodine
    E. alcohol.

In questions 11–15, pair the following causative organisms with the diseases:

    A. bacteria B. virus C. nematode D. protozoon E. fluke.

11. common cold 12. tuberculosis 13. malaria.
14. diphtheria 15. cholera.

In questions 16–20 pair the following parts of the digestive system with their function.

    A. stomach B. duodenum C. ileum D. pancreas E. colon.

16. absorption of water 17. makes food acid
18. absorbs digested food 19. contains alkaline food
20. mixes food into chyme.

Questions 21–30. Each question contains four statements. If (a), (b) and (c) only are correct answer A; (a) and (c) only are correct answer B, (b) and (d) only are correct answer C; (a) only correct answer D; any other combination correct answer E.

21. Tooth decay can be reduced by:

    (a) frequent brushing of the teeth
    (b) eating a balanced diet
    (c) not eating between meals
    (d) eating plenty of sugar.

22. Which of the following elements are necessary in the diet?

    (a) lead (b) calcium (c) tin (d) iron.

23. Respiration always involves:

    (a) the production of energy
    (b) carbohydrates
    (c) oxygen
    (d) production of water.

24. A young baby needs:

    (a) plenty of sleep (b) to be kept warm
    (c) comfortable clothing (d) a balanced diet.

25. Two individuals that look alike can:

    (a) have the same genotype (b) have the same phenotype
    (c) be homozygous (d) be heterozygous.

26. The skin:

    (a) helps to control body temperature
    (b) controls water loss from the body

(c) contains sensory cells
(d) is a dead layer covering the body.

27. A neurone (nerve cell) contains:

    (a) a nucleus   (b) chloroplasts
    (c) mitochondria   (d) a cell wall.

28. The chief substances in bone are:

    (a) iron   (b) calcium   (c) vitamin D   (d) phosphorus.

29. The liver:

    (a) produces bile   (b) stores protein
    (c) stores carbohydrates   (d) stores water.

30. Red blood cells:

    (a) are called erythrocytes
    (b) will swell if put into pure water
    (c) transport oxygen
    (d) ingest bacteria.

Questions 31–40 consist of two statements. If both statements are true, and the second follows from the first, answer A; if both statements are true, but the second does not follow from the first, answer B; if the first is true and the second false, answer C; if the first is false and the second true, answer D; if both are false answer E.

31. Carnivores are the end consumers in a food web	because	Carnivores eat herbivores or other carnivores.
32. Arteries carry blood to the heart	because	Arteries are thick-walled.
33. Serum and lymph have the same composition	because	Serum and lymph are produced by lympho-cytes.
34. Carbohydrates have a higher calorific value than fats	because	Carbohydrates and fats both contain carbon, hydrogen and oxygen.
35. Cavity walls insulate against heat loss	because	Cavity walls have an air space between the two layers.
36. Cholera is caused by a bacterium	because	Cholera is a contagious disease.
37. Thyroxine affects the rate of metabolism	because	Thyroxine is a hormone.
38. The cochlea is the organ of balance	because	The cochlea contains the organ of Corti.

39. The retina is the light-   because   The retina contains rods
    sensitive part of the   and cones.
    eye

40. Fleas do not transmit   because   Fleas feed on human
    diseases   hair.

ANSWERS TO MULTIPLE-CHOICE QUESTIONS

1.	B	11.	B	21.	A	31.	A
2.	C	12.	A	22.	C	32.	B
3.	E	13.	D	23.	D	33.	E
4.	A	14.	A	24.	E	34.	D
5.	D	15.	A	25.	E	35.	A
6.	A	16.	E	26.	A	36.	C
7.	B	17.	A	27.	B	37.	B
8.	C	18.	C	28.	C	38.	D
9.	E	19.	B	29.	B	39.	A
10.	B	20.	A	30.	A	40.	E.

IMPORTANT PEOPLE IN THE HISTORY OF HUMAN BIOLOGY

Banting and Best: two Canadian physiologists, working in Toronto under Professor Macleod, first produced insulin extract in 1921. The first diabetes patient was treated in 1922.

Bayliss and Starling: the first people to isolate a hormone, secretin.

Braille (1809–1853): developed an alphabet of forty-three symbols which could be represented as indentations on a card. This enabled blind people to read.

Calmette and Guerin: two Frenchmen who first prepared the B.C.G. (Bacillus Calmette–Guerin) vaccine against tuberculosis. It has been used to vaccinate school children since 1953.

Chadwick (1800–1890): a social reformer, greatly concerned with the living conditions of the poor. He was one of the instigators of the Public Health Act of 1848.

Curie, Pierre (1859–1906) *and Marie* (1867–1934): Isolated the elements radium and plutonium (1898). Their research on radioactivity was the basis of the radio-therapy of cancer.

Darwin, Charles (1809–1892): after extensive observations he formulated the theory of natural selection being the basis of evolution.

Domagk (1895–1964): first discovered the effects of sulphanilamide drugs on the toxins produced by bacteria in the body.

Eijkman (1858–1930): the first to show that diseases could result from the absence of substances in the body. (See *Hopkins*.)

Fleming (1881–1955): discovered that a substance from the fungus *Penicillium* killed the bacterium *Staphylococcus* (1928). This was the first antibiotic to be discovered.

Hopkins, Gowland (1861–1947): he pioneered the research into the necessity for vitamins in the diet.

Iwanowski: first discovered (in 1892) that particles smaller than bacteria could infect plants. These were later called viruses.

Jenner (1749–1823): first to introduce vaccination against small-pox (1796).

Koch (1843–1910): traced the life-history of the anthrax bacillus, and discovered the tuberculosis bacterium (1882).

Lamark (1744–1829): advanced a theory of evolution based on the inheritance of acquired characteristics.

Leeuwenhoek (1632–1723): an early microscopist who first discovered bacteria, described the structure of red blood cells and
observed blood flowing through capillaries.

Lister (1827–1912): first (1865) person to use antiseptic (carbolic)
during surgery, and greatly reduced the death rate from post-
operative septicaemia.

Manson (1844–1922): one of the founders of the London School of
Tropical Medicine. He advanced the theory that malaria was
transmitted by a mosquito.

Mendel (1822–1884): the father of modern genetics. Using pea
plants he showed that one character is dominant to another, and
that different characters are inherited independently.

Pasteur (1822–1898): from observations on fermentation, he put
forward the germ theory of disease, *i.e.* that diseases are caused
by microbes from the air, and not spontaneously generated
from non-living substances.

Pavlov (1849–1936): first demonstrated the conditioned reflex.

Röntgen (1845–1923): discovered X-rays (1895) which are used in
the diagnosis of disease (especially tuberculosis) and for examin-
ing internal damage, *e.g.* broken bones.

Ross (1857–1932): Proved that the mosquito *Anopheles* is the
vector of malaria.

Stanley (1904–   ): The first to isolate virus particles (1935).

GLOSSARY OF TERMS

$\overset{\circ}{A}$: angstrom unit $= 10^{-10}$ metres.

acetyl choline: the substance which brings about the passage of a nerve impulse across a synapse.

active transport: the accumulation of substances within a cell against the concentration gradient.

aerobic respiration: cellular respiration which involves oxygen in the breakdown of the substrate.

alveoli: the fine air sacs making up the respiratory surface of the lung.

amino acid: an organic acid which always contains nitrogen. One of the basic units of protein synthesis.

amnion: a membrane which forms a liquid-filled sac around the foetus in the uterus.

amylase: an enzyme which digests starch.

anaerobic respiration: cellular respiration which does not involve oxygen in the breakdown of the substrate.

antibiotic: a substance released by one micro-organism, *e.g.* a fungus, which kills other micro-organisms, *e.g.* bacteria.

antibody: a substance present in the blood serum, produced in response to the presence of foreign protein in the blood.

antigen: a foreign substance, usually protein (*e.g.* bacteria, red blood cells), present in the blood and stimulating the production of antibodies.

antitoxin: a substance produced in the blood stream which combines with a toxin to prevent its activity.

artery: a blood vessel carrying blood away from the heart.

ATP: adenosine triphosphate. The chemical involved in transporting the energy produced during respiration to various sites in a cell.

autosome: a normal chromosome, *i.e.* not a sex chromosome.

biconcave: curved inwards on both sides.

blastocyte: the ball of cells produced early in the growth of an embryo.

bolus: = ball. A soft mass of chewed food.

buccal cavity: the part of the mouth behind the lips, up to the pharynx.

callus: an abnormal growth of cells to produce a hard or thickened tissue.

calorie: the amount of heat required to raise 1 g of water through 1°C. 1 calorie = 4·2 joules. The kilocalorie used to measure the energy available from food is 1000 calories.

calorific value: the energy available from a stated mass of food, usually 1 g. It is quoted in kilocalories.

capillary: a very fine blood vessel.

carbohydrate: *e.g.* sugars, starch. A compound containing the elements carbon, hydrogen and oxygen only, with twice as many hydrogen atoms as oxygen atoms. Those with up to six carbon atoms are called *monosaccharides*, *e.g.* glucose ($C_6H_{12}O_6$); those with six–twelve carbon atoms are *disaccharides*, *e.g.* sucrose ($C_{12}H_{22}O_{11}$): and those with a larger number of carbon atoms are *polysaccharides*, *e.g.* starch and glycogen. All carbohydrates are digested to monosaccharides before they are absorbed.

cardiac: belonging to the heart.

carrier: a person who is infected with a disease organism, but who does not show the symptoms. In humans, a female who is heterozygous for a sex-linked gene.

cartilage: a tough elastic tissue for support and shock-absorption.

centriole: a small body near the nuclear membrane of a cell.

centromere: the part of a chromosome which becomes attached to the spindle during cell division.

chlorophyll: the green pigment in plants. It is essential for photosynthesis.

choline esterase: an enzyme which destroys acetyl choline at a synapse, thus terminating the passage of a nerve impulse.

chromatid: one of the threads into which a chromosome splits during cell division.

chromosome: one of a pair of deeply staining rod-like bodies found in the nucleus of a cell. The chromosomes carry the genetic information, and are of constant number in the cells of a particular species. The members of each identical pair of chromosomes in a nucleus are *homologous*. Homologous chromosomes become paired during cell division.

cleavage: the early divisions of the cells of a zygote, before the blastocyte is formed.

cranial nerves: nerves running directly from the brain to the organ they supply, *i.e.* they do not run through the spinal cord.

cytoplasm: the protoplasm in a cell, excluding the nucleus.

dental formula: a method of describing the number and type of teeth.

dentine: the hard material making up most of a tooth. It is covered by a harder enamel.

differentiate: (referring to cells) becoming specialised to perform a particular function.

dilate: to become larger in diameter.

DNA: deoxyribose nucleic acid. The substance from which genes are made.

effector: an organ, *e.g.* a muscle, which responds to a nerve impulse.

emulsify: to break down into fine droplets. For example oil, which does not mix with water, can be made to stay suspended in water if it is emulsified.

enamel: the hard layer covering a tooth.

endemic: throughout an area, or within a population in an area; frequently referred to a disease which is found widely throughout a population during a long period of time.

endo-: within.

endoplasmic reticulum: a system of membranes running throughout the cytoplasm of a cell.

end plate: the ending of a motor nerve in an effector.

enzyme: a protein which accelerates a chemical reaction. They are very specific and sensitive to changes in the environment, *e.g.* heat or acidity.

epiglottis: a flap of muscular tissue closing off the top of the trachea.

erythrocyte: a red blood cell.

exo-: outside.

fat: substances containing carbon, hydrogen and oxygen, but containing proportionally more oxygen than carbohydrates. They have a higher calorific value than carbohydrates, and are important solvents for some vitamins.

ganglion: a collection of nerve cell bodies, and possibly nerve fibres.

gastric: related to the stomach.

gene: the individual units, probably of DNA, storing the genetic information in a cell. They are arranged in a row on chromosomes.

genotype: the genetic make-up of an individual.

glycogen: a carbohydrate, related to starch. A storage product produced by the liver.

haemoglobin: the red pigment in red blood cells. It combines readily with oxygen to form oxyhaemoglobin, and so is able to transport oxygen from the lungs to the tissues.

hepatic: related to the liver.

homeostatic: keeping the internal environment, *e.g.* temperature, constant.

homo-: the same.

hormone: the secretion of a ductless (endocrine) gland which is carried in the blood stream, and affects an organ some distance from the gland.

ingest: to take in.

ion: an atom or molecule which has an electric charge on it.

lacteals: fine lymphatic vessels running into the villi of the intestine

leucocytes: white blood cells.

lipase: an enzyme which digests fats.

lymph: a clear liquid circulating in the lymphatic system. It is similar to, but not identical with, blood plasma. *Perilymph* and *endolymph* are liquids in the inner ear not connected with the lymphatic system.

lymphocyte: a form of leucocyte produced in the lymph glands.

matrix: non-living material laid down by living cells which are embedded in the matrix, *e.g.* the ground material of bone or cartilage.

metabolic water: water which is produced during the respiration of hydrogen-containing compounds, *e.g.* glucose.

mitochondria: rod-like inclusions in the protoplasm. They contain, particularly, the enzymes of respiration.

μ: the Greek letter mu. Equals 10^{-7} metres.

neurone: a nerve cell. One of the units which make up the nervous system.

nitrogen bases: organic chemicals with NH_2 group(s) in them.

nucleolus: a spherical body within the nucleus.

nucleus: the part of a cell, bounded by a membrane, which contains the genetic information, and controls the activity of the cell.

obesity: putting on too much weight.

olfactory: concerned with smelling.

organ: a collection of specialized cells which perform a particular function, or a limited number of functions.

osmoregulation: the control of the amount of water in the body, and of the concentrations of dissolved substances.

osmosis: the passage of a solvent from a dilute solution to a more concentrated one through a semipermeable membrane.

ossification: becoming impregnated with calcium and other salts to form bone.

osteocyte: a bone-secreting cell.

ovum: an egg.

oxygen deficit: the condition which arises when cells, which normally respire aerobically, switch to anaerobic respiration, *e.g.* during intense exercise. The products of anaerobic respiration accumulate in the cells, and, after the activity has stopped these are removed by oxidation. Thus after activity, the cells require more oxygen than they would normally.

pandemic: spreading rapidly throughout a population, and then dying out.

penta-: five. For example pentadactyl means five-fingered.

peri-: around.

pericardium: a tough membrane surrounding the heart.

peripheral: around the outside, *e.g.* peripheral blood vessels are those in the skin.

peristalsis: contraction of the circular muscles in a tubular organ, *e.g.* the gut.

phagocyte: a migrating white blood cell which ingests bacteria.

phenotype: what a person looks like. The phenotype is produced by the interaction of the genotype and the environment.

plasma: the liquid part of the blood. It contains dissolved salts, proteins and other organic substances.

plasma membrane: the fine membrane surrounding the protoplasm of a cell.

pleural: literally = side, but refers to the lining of the chest cavity.

portal system: a blood vessel joining one organ to another, *i.e.* the blood from it does not flow directly into the blood circulatory system.

protease: an enzyme which breaks down proteins.

protein: substances with very high molecular weights (relative molecular masses) built up from amino acids (*q.v.*).

protoplasm: the living substance of a cell.

pulmonary: concerned with the lungs.

receptor: part of the nervous system adapted to receive stimuli, *e.g.* a touch cell or the eye.

ribosome: small particles in the cytoplasm concerned with synthesising proteins.

serum: the straw-coloured liquid which is left when blood clots. It is similar to plasma, but contains less soluble protein.

spermatozoon: a male gamete.

sphincter: a ring of muscle around a tubular organ. The contraction of these muscles closes off the organ.

spp.: the abbreviation for species.

stereoscopic: being able to appreciate the depth of an object or view.

subcutaneous: below the skin.

synovial membrane: a membrane inside tendon sheaths and lining the capsules of movable joints.

tissue: a collection of one type of cell. Several different types of tissue may contribute to making up an organ.

tone: the inherent tension in living muscles.

toxin: a poison produced by a plant or animal.

trace element: an element needed in very small amounts for the health of the body.

urea: an organic compound containing carbon, hydrogen, oxygen, and nitrogen. It is produced in the liver during the breakdown of amino acids, and is a component of urine.

vacuole: a liquid-filled sac within the protoplasm.

vector: an animal (usually an insect) which carries a disease organism from one host to another.

vein: a vessel carrying blood to the heart.

villus: a fine finger-like process increasing the surface area of a structure.

vitamin: organic compounds needed in small amounts for the healthy functioning of the body.

zygote: the single cell produced by the fusion of a male and a female gamete.

INDEX